PETROCHEMICAL PRODUCTION PROCESS

PETROCHEMICAL PRODUCTION PROCESS

R.K. Singh

RANDOM PUBLICATIONS

NEW DELHI - 110 002 (INDIA)

butadiene (SBR) rubber, styrene-butadiene latex, SIS (styrene-isoprene-styrene), S-EB-S (styrene-ethylene/butylene-styrene), styrene-divinylbenzene (S-DVB), styrene-acrylonitrile resin (SAN) and unsaturated polyesters.

These materials are used in rubber, plastic, insulation, fiberglass, pipes, automobile and boat parts, food containers, and carpet backing.

Production

Styrene is produced in industrial quantities from ethylbenzene, which is in turn prepared on a large scale by alkylation of benzene with ethylene.

Dehydrogenation of Ethylbenzene

Styrene is most commonly produced by the catalytic dehydrogenation of ethylbenzene. Ethylbenzene is mixed in the gas phase with 10–15 times its volume in high-temperature steam, and passed over a solid catalyst bed. Most ethylbenzene dehydrogenation catalysts are based on iron(III) oxide, promoted by several percent potassium oxide or potassium carbonate.

Steam serves several roles in this reaction. It is the source of heat for powering the endothermic reaction, and it removes coke that tends to form on the iron oxide catalyst through the water gas shift reaction. The potassium promoter enhances this decoking reaction. The steam also dilutes the reactant and products, shifting the position of chemical equilibrium towards products.

A typical styrene plant consists of two or three reactors in series, which operate under vacuum to enhance the conversion and selectivity. Typical per-pass conversions are ca. 65% for two reactors and 70-75% for three reactors. Selectivity to styrene is 93-97%. The main byproducts are benzene and toluene.

Because styrene and ethylbenzene have similar boiling points (145 and 136 °C, respectively), their separation requires tall distillation towers and high return/reflux ratios. At its distillation temperatures, styrene tends to polymerize.

To minimize this problem, early styrene plants added elemental sulfur to inhibit the polymerization. During the 1970s, new free radical inhibitors consisting of nitrated phenol-based retarders were developed. More recently, a number of additives have been developed that exhibit superior inhibition against polymerization. However, the nitrated phenols are still widely used because of their relatively low cost. These reagents are added prior to the distillation.

Improving conversion and so reducing the amount of ethylbenzene that must be separated is the chief impetus for researching alternative routes to styrene. Other than the POSM process, none of these routes like obtaining styrene from butadiene have been commercially demonstrated.

Via Ethylbenzenehydroperoxide

Commercially styrene is also co-produced with propylene oxide in a process known as POSM (Lyondell Chemical Company) or SM/PO (Shell) for styrene monomer/propylene oxide. In this process ethylbenzene is treated with oxygen to form the ethylbenzene hydroperoxide. This hydroperoxide is then used to oxidize propylene to propylene oxide.

The resulting 2-phenylethanol is dehydrated to give styrene:

$$C_6H_5CH_2CH_3 + O_2 \rightarrow C_6H_5CH_2CH_2O_2H$$

$$C_6H_5CH_2CH_2O_2H + CH_3CH{=}CH_2 \rightarrow C_6H_5CH_2CH_2OH + CH_3CHCH_2O$$

$$C_6H_5CH_2CH_2OH \rightarrow C_6H_5CH{=}CH_2 + H_2O$$

Laboratory Synthesis

A laboratory synthesis of styrene entails the decarboxylation of cinnamic acid. Styrene was first prepared by this method.

Other Methods

Styrene can be produced from toluene and methanol, which are cheaper raw materials than those in the conventional process. Historically, however, this process has suffered from low selectivity due to competing decomposition of methanol.

Exelus Inc. claims to have developed this process with commercially viable selectivities, at 400-425 °C and atmospheric pressure, by forcing these components through a proprietary zeolitic catalyst. It is reported that an approximately 9:1 mixture of styrene and ethylbenzene is obtained, with a total styrene yield of over 60%.

Another developing route to styrene is via benzene and ethane. This process is being developed by Snamprogetti S.p.A. and Dow. Ethane, along with ethylbenzene, is fed to a dehydrogenation reactor with a catalyst capable of simultaneously producing styrene and ethylene. The dehydrogenation effluent is cooled and separated and the ethylene stream is recycled to the alkylation unit. The process attempts to overcome previous shortcomings in earlier attempts to develop production of styrene from ethane and benzene, such as inefficient recovery of aromatics, production of high levels of heavies and tars, and inefficient separation of hydrogen and ethane. Development of the process is ongoing.

Health Effects

Styrene is regarded as a "hazardous chemical", especially in case of eye contact, but also in case of skin contact, of ingestion and of inhalation, according to several sources. The US EPA has described styrene to be "a suspected toxin to the gastrointestinal tract, kidney, and respiratory system, among others."

On 10 June 2011, the US National Toxicology Program has described styrene as "reasonably anticipated to be a human carcinogen". However, an academic panel, funded by the styrene industry, recently reviewed the relevant scientific literature and concluded that "The available epidemiologic evidence does not support a causal relationship between styrene exposure and any type of human cancer."

The U.S. EPA does not have a cancer classification for styrene, but currently is evaluating styrene's cancer-causing potential through its Integrated Risk Information System (IRIS) program.

The U.S. National Toxicology Program of the U.S. Department of Health and Human Services also currently is evaluating styrene's potential toxicity To date, no regulatory body anywhere in the world has classified styrene as a known human carcinogen, although several refer to it in various contexts as a possible or potential human carcinogen. The International Agency for Research on Cancer considers styrene to be "possibly carcinogenic to humans.". Chronic exposure to styrene leads to tiredness/lethargy, memory deficits, headaches and vertigo.

According to the Styrene Information and Research Centre (an organization representing nearly all of the "North American styrene industry"), polystyrene plastic neither contains, nor breaks down into bisphenol A (BPA), a chemical used in plastic compounds that leads to developmental and reproductive problems in both adults and children.

Polystyrene

Polystyrene also known as Thermocole, abbreviated following ISO Standard PS, is an aromatic polymer made from the monomer styrene, a liquid hydrocarbon that is manufactured from petroleum by the chemical industry. Polystyrene is one of the most widely used plastics, the scale being several billion kilograms per year.

Polystyrene is a thermoplastic substance, which is in solid (glassy) state at room temperature, but flows if heated above its glass transition temperature of about 100 °C (for molding or extrusion), and becomes solid again when cooled. Pure solid polystyrene is a colorless, hard plastic with limited flexibility. It can be cast into molds with fine detail. Polystyrene can be transparent or can be made to take on various colours.

Solid polystyrene is used, for example, in disposable cutlery, plastic models, CD and DVD cases, and smoke detector housings. Products made from foamed polystyrene are nearly ubiquitous, for example packing materials, insulation, and foam drink cups.

Polystyrene can be recycled, and has the number "6" as its recycling symbol. The increasing oil prices have increased the value of polystyrene for recycling. No known microorganism has yet been shown to biodegrade polystyrene, and it is often abundant as a form of pollution in the outdoor environment, particularly along shores and waterways especially in its low density cellular form.

History

Polystyrene was discovered in 1839 by Eduard Simon, an apothecary in Berlin. From storax, the resin of the Turkish sweetgum tree Liquidambar orientalis, he distilled an oily substance, a monomer which he named styrol. Several days later, Simon found that the styrol had thickened, presumably from oxidation, into a jelly he dubbed styrol oxide ("Styroloxyd"). By 1845 English chemist John Blyth and German chemist August Wilhelm von Hofmann showed that the same transformation of styrol took place in the absence of oxygen. They called their substance metastyrol.

Analysis later showed that it was chemically identical to Styroloxyd. In 1866 Marcelin Berthelot correctly identified the formation of metastyrol from styrol as a polymerization process. About 80 years went by before it was realized that heating of styrol starts a chain reaction which produces macromolecules, following the thesis of German organic chemist Hermann Staudinger (1881–1965). This eventually led to the substance receiving its present name, polystyrene.

The company I. G. Farben began manufacturing polystyrene in Ludwigshafen, Germany, about 1931, hoping it would be a suitable replacement for die-cast zinc in many applications. Success was achieved when they developed a reactor vessel that extruded polystyrene through a heated tube and cutter, producing polystyrene in pellet form. Before 1949, the chemical engineer Fritz Stastny (1908–

1985) developed pre-expanded PS beads by incorporating aliphatic hydrocarbons, such as pentane. These beads are the raw material for moulding parts or extruding sheets. BASF and Stastny applied for a patent which was issued in 1949. The moulding process was demonstrated at the Kunststoff Messe 1952 in Düsseldorf. Products were named Styropor.

The crystal structure of isotactice polystyrene was reported by Giulio Natta.

In 1959, the Koppers Company in Pittsburgh, Pennsylvania, developed expanded polystyrene (EPS) foam.

Structure

In chemical terms, polystyrene is a long chain hydrocarbon wherein alternating carbon centers are attached to phenyl groups (the name given to the aromatic ring benzene. Polystyrene's chemical formula is (C8H8)n; it contains the chemical elements carbon and hydrogen.

The materials properties are determined by short range van der Waals attractions between polymers chains. Since the molecules are long hydrocarbon chains that consist of thousands of atoms, the total attractive force between the molecules is large.

When heated (or, equivalently, deformed at a rapid rate, due to a combination of viscoelastic and thermal insulation properties), the chains are able to take on a higher degree of conformation and slide past each other. This intermolecular weakness (versus the high intramolecular strength due to the hydrocarbon backbone) confers flexibility and elasticity. The ability of the system to be readily deformed above its glass transition temperature allows polystyrene (and thermoplastic polymers in general) to be readily softened and molded upon heating.

Polymerization

Polystyrene results when styrene monomers interconnect. In the polymerization one carbon-carbon double bond (in the vinyl group) is replaced by a much stronger carbon-carbon single bond, hence it is very difficult to depolymerize polystyrene. Typically about a few thousand monomers comprise a chain of polystyrene, giving a molecular weight of 100,000–400,000.

A 3-D model would show that each of the chiral backbone carbons lies at the centre of a tetrahedron, with its 4 bonds pointing toward the vertices. Say, the -C-C- bonds are rotated so that the backbone chain lies entirely in the plane of the diagram. From this flat schematic, it is not evident which of the phenyl (benzene) groups are angled toward us from the plane of the diagram, and which ones are angled away. The isomer where all of them are on the same side is called isotactic polystyrene, which is not produced commercially.

Atactic Polystyrene

The only commercially important form of polystyrene is atactic, which means that the phenyl groups are randomly distributed on both sides of the polymer chain. This random positioning prevents the chains from ever aligning with sufficient regularity to achieve any crystallinity. The plastic has a glass transition temperature Tg of ~90C. Polymerization is initiated with free radicals.

Isotactic and Syndiotactic Polystyrene

Ziegler-Natta polymerization can produce an ordered syndiotactic polystyrene with the phenyl groups positioned on alternating sides of the hydrocarbon backbone. This form is highly crystalline with a Tm of 270 °C (518 °F). Such materials are not commercially produced because the polymerization is slow.

Extruded polystyrene is about as strong as an unalloyed aluminium, but much more flexible and much lighter (1.05 g/cm3 vs. 2.70 g/cm3 for aluminium).

Degradation

Because it is an aromatic hydrocarbon, it burns with an orange-yellow flame, giving off soot, as is characteristic of materials containing aromatic rings. Complete oxidation of polystyrene produces carbon dioxide and water vapour. Because of its chemical inertness, polystyrene is used to fabricate containers for chemicals, solvents, and foods.

Forms Produced

Properties	
Density	1.05 g/cm3
Density of EPS	16–640 kg/m3
Dielectric constant	2.4–2.7
Electrical conductivity (s)	10"16 S/m
Thermal conductivity (k)	0.08 W/(m·K)
Young's modulus (E)	3000–3600 MPa
Tensile strength (st)	46–60 MPa
Elongation at break	3–4%
Notch test	2–5 kJ/m2

Contd...

Properties	
Glass transition temperature	95 °C
Melting point	240 °C
Vicat B	90 °C
Linear expansion coefficient (a)	8×10"5/K
Specific heat (c)	1.3 kJ/(kg·K)
Water absorption (ASTM)	0.03–0.1
Decomposition	X years, still decaying

Polystyrene is commonly injection molded or extruded, while expanded polystyrene is either extruded or molded in a special process. Polystyrene copolymers are also produced; these contain one or more other monomers in addition to styrene. In recent years the expanded polystyrene composites with cellulose and starch have also been produced.

Extruded closed-cell polystyrene foam is sold under the trademark Styrofoam by Dow Chemical. This term is often used informally for other foamed polystyrene products.

Polystyrene is used in some polymer-bonded explosives:

Polystyrene PBX examples Name	*Explosive ingredients*	*Binder ingredients*
PBX-9205	RDX 92%	Polystyrene 6%; DOP 2%
PBX-9007	RDX 90%	Polystyrene 9.1%; DOP 0.5%; resin 0.4%

Sheet or Molded Polystyrene

CD case made from general purpose polystyrene (GPPS) and high impact polystyrene (HIPS).

Disposable Polystyrene Razor

Polystyrene (PS) is economical, and is used for producing plastic model assembly kits, plastic cutlery, CD "jewel" cases, smoke detector housings, license plate frames, and many other objects where a fairly rigid, economical plastic is desired. Production methods include thermoforming and injection molding.

Polystyrene Petri dishes and other laboratory containers such as test tubes and microplates play an important role in biomedical research and science. For these uses, articles are almost always made by injection molding, and often sterilized post-molding, either by irradiation or treatment with ethylene oxide. Post-mold surface modification, usually with oxygen-rich plasmas, is often done to introduce polar groups. Much of modern biomedical research relies on the use of such products; they therefore play a critical role in pharmaceutical research.

Foams

Polystyrene foams are good thermal insulators and are therefore often used as building insulation materials, such as in structural insulated panel building systems. They are also used for non-weight-bearing architectural structures (such as ornamental pillars). PS foams exhibit also good damping properties, therefore it is used widely in packaging.

Expanded Polystyrene

Expanded polystyrene (EPS) is a rigid and tough, closed-cell foam. It is usually white and made of pre-expanded polystyrene beads. Familiar uses include molded sheets for building insulation and packing material ("peanuts") for cushioning fragile items inside boxes. Sheets are commonly packaged as rigid panels (size 4 by 8 or 2 by 8 feet in the United States), which are also known as "bead-board".

Thermal resistivity is usually about 28 m·K/W (or R-4 per inch). Some EPS boards have a flame spread of less than 25 and a smoke-developed index of less than 450, which means they can be used without a fire barrier (but require a 15 minute thermal barrier) according to US building codes. A growing use of EPS in construction is insulating concrete forms. The density range is about 16–640 kg/m3. The most common processing method is thermal cutting with hot wires.

Extruded Polystyrene Foam

Extruded polystyrene foam (XPS) consists of closed cells, offers improved surface roughness and higher stiffness and reduced thermal conductivity. The density range is about 28 – 45 kg/m3.

Extruded polystyrene material is also used in crafts and model building, particularly architectural models. Because of the extrusion manufacturing process, XPS does not require facers to maintain its thermal or physical property performance.

Thus, it makes a more uniform substitute for corrugated cardboard. Thermal resistivity is usually about 35 m·K/W (or R-5 per inch in American customary units).Styrofoam is a trademarked name for XPS; however, it is often also used in the United States as a generic name for all polystyrene foams.

Copolymers

Pure polystyrene is brittle, but hard enough that a fairly high-performance product can be made by giving it some of the properties of a stretchier material, such as polybutadiene rubber.

The two such materials can never normally be mixed because of the amplified effect of intermolecular forces on polymer insolubility, but if polybutadiene is added during polymerization it can become chemically bonded to the polystyrene, forming a graft copolymer which helps to incorporate normal polybutadiene into the final mix, resulting in high-impact polystyrene or HIPS, often called "high-impact plastic" in advertisements.

One commercial name for HIPS is Bextrene. Common applications of HIPS include toys and product casings. HIPS is usually injection molded in production. Autoclaving polystyrene can compress and harden the material.

Several other copolymers are also used with styrene. Acrylonitrile butadiene styrene or ABS plastic is similar to HIPS: a copolymer of acrylonitrile and styrene, toughened with polybutadiene. Most electronics cases are made of this form of polystyrene, as are many sewer pipes. SAN is a copolymer of styrene with acrylonitrile, and SMA one with maleic anhydride. Styrene can be copolymerized with other monomers; for example, divinylbenzene can be used for cross-linking the polystyrene chains to give the polymer used in Solid phase peptide synthesis.

Oriented Polystyrene

Oriented polystyrene (OPS) is produced by stretching extruded PS film, improving visibility through the material by reducing haziness and increasing stiffness. This is often used in packaging where the manufacturer would like the consumer to see the enclosed product. Some benefits to OPS are that it is less expensive to produce than other clear plastics such as PP, PET and HIPS, and it is less hazy than HIPS or PP. The main disadvantage to OPS is that it's brittle. It will crack or tear easily.

Disposal and Environmental Issues

Polystyrene is easily recycled. Due its light weight (especially if foamed) it is not economical to collect in its original form. However if the waste material goes through an initial compaction process the material changes density from typically 30 kg/m3 to 330 kg/m3 and becomes a recyclable commodity of high value for producers of recycled plastic pellets. It is generally not accepted in curbside collection recycling programs. In Germany polystyrene is collected, as a consequence of the packaging law (Verpackungsverordnung) that requires manufacturers to take responsibility for recycling or disposing of any packaging material they sell.

In the US and many other countries the interest in recycling polystyrene has led to collection points being established. The producers of large quantities of polystyrene waste (50 tons per year or more) who have invested in the EPS compactors are able to sell the compacted blocks to plastic recyclers.

Environmental Impact

Discarded polystyrene does not biodegrade for hundreds of years and is resistant to photolysis. Because of this stability, very little of the waste discarded in today's modern, highly engineered landfills biodegrades. Because degradation of materials creates potentially harmful liquid and gaseous by-products that could contaminate groundwater and air, today's landfills are designed to minimize contact with air and water required for degradation, thereby practically eliminating the degradation of waste.

Polystyrene foam is a major component of plastic debris in the ocean, where it becomes toxic to marine life. Foamed polystyrene blows in the wind and floats on water, and is abundant in the outdoor environment. Polystyrene foams are produced using blowing agents that form bubbles and expand the foam.

In expanded polystyrene, these are usually hydrocarbons such as pentane, which may pose a flammability hazard in manufacturing or storage of newly manufactured material, but have relatively mild environmental impact.

However, extruded polystyrene is usually made with hydrochlorofluorocarbons (HCFC) blowing agents which have effects on ozone depletion and on global warming. Their ozone depletion potential is greatly reduced relative to chlorofluorocarbon (CFC) which were formerly used, but their global warming potential can be on the order of 1000 or more, meaning it has 1000 times greater effect on global warming than does carbon dioxide.

In 2007, Eben Bayer, a student at Rensselaer Polytechnic Institute, invented an environmentally-friendly replacement for polystyrene packaging using the mineral perlite bonded with lignin-containing agricultural waste, degraded by fungi, which he calls Ecocradle.

Recycling

The resin identification code symbol for polystyrene Currently, most polystyrene products are not recycled due to the lack of incentive to invest in the compactors and logistical systems required.

Expanded polystyrene scrap can be easily added to products such as EPS insulation sheets and other EPS materials for construction applications. Commonly, manufacturers cannot obtain sufficient scrap because of the aforementioned collection issues.

When it is not used to make more EPS, foam scrap can be turned into clothes hangers, park benches, flower pots, toys, rulers, stapler bodies, seedling containers, picture frames, and architectural molding from recycled PS.

Recycled EPS is also used in many metal casting operations. Rastra is made from EPS that is combined with cement to be used as an insulating amendment in the making of concrete foundations and walls. American manufacturers have produced insulating concrete forms made with approximately 80% recycled EPS since 1993. However, polystyrene recycling is not a closed loop, producing more polystyrene; polystyrene cups and other packaging materials are instead usually used as fillers in other plastics, or in other items that cannot themselves be recycled and are thrown away.

Incineration

If polystyrene is properly incinerated at high temperatures, the chemicals generated are water, carbon dioxide, a complex mixture of volatile compounds, and carbon soot. According to the American Chemistry Council, when polystyrene is incinerated in modern facilities, the final volume is 1% of the starting volume; most of the polystyrene is converted into carbon dioxide, water vapour, and heat. Because of the amount of heat released, it is sometimes used as a power source for steam or electricity generation.

When polystyrene was burned at temperatures of 800–900 °C (the typical range of a modern incinerator), the products of combustion consisted of "a complex mixture of polycyclic aromatic hydrocarbons (PAHs) from alkyl benzenes to benzoperylene. Over 90 different compounds were identified in combustion effluents from polystyrene."

When burned without enough oxygen or at lower temperatures (as in a campfire or a household fireplace), polystyrene can produce polycyclic aromatic hydrocarbons, carbon black, and carbon monoxide, as well as styrene monomers.

Burial

This section does not cite any references or sources. Please help improve this section by adding citations to reliable sources. Unsourced material may be challenged and removed.

The neutrality of this article is disputed.

Foam cups and other polystyrene products can be safely buried in landfills, since it is as stable as concrete or brick. No plastic film is required to protect the air and underground water.

Reducing

Some effort is being made to find alternatives to polystyrene foam, especially in restaurant settings. Restricting the use of foamed polystyrene takeout food packaging is a priority of many solid waste environmental organizations. However, the Plastics Foodservice Packaging Group counters that in US, less than 1% by weight of solid waste disposed is polystyrene.

A campaign to achieve the first ban of polystyrene foam from the food & beverage industry in Canada was launched in Toronto as of January 2007, by local non-profit organization NaturoPack. Portland, Ore., and San Francisco are among about one hundred cities in the United States that currently have some sort of ban on polystyrene foam in restaurants. For instance, in 2007 restaurants in Oakland, California were required to switch to disposable food containers that will biodegrade if added to food compost.

Although polystyrene can be recycled at recycling facilities, most polystyrene is not recycled. The EPA (United States Environmental Protection Agency) estimates that 25 billion polystyrene cups are tossed every year. Since polystyrene degrades very slowly- more than 500 years for a single cup– the EPA considers this a serious environmental problem. Several green leaders, from the Dutch Ministry of the Environment to Starbucks' Green Team, advise that individuals reduce their environmental impact by using reusable coffee cups.

Finishing

This section does not cite any references or sources. Please help improve this section by adding citations to reliable sources. Unsourced material may be challenged and removed. In the United States, environmental protection regulations prohibit the use of solvents on polystyrene (which would dissolve the polystyrene and de-foam most foams anyway).

Some acceptable finishing materials are:

- Water-based paint (artists have created paintings on polystyrene with gouache)
- Mortar or acrylic/cement render, often used in the building industry as a weather-hard overcoat that hides the foam completely after finishing the objects.
- Cotton wool or other fabrics used in conjunction with a stapling implement.

Safety—Health

Based on scientific tests over five decades, government safety agencies have determined that polystyrene is safe for use in foodservice products. For example, polystyrene meets the stringent standards of the U.S. Food and Drug Administration and the European Commission/ European Food Safety Authority for use in packaging to store and serve food. The Hong Kong Food and Environmental Hygiene Department recently reviewed the safety of serving various foods in polystyrene foodservice products and reached the same conclusion as the U.S. FDA.

From 1999 to 2002, a comprehensive review of the potential health risks associated with exposure to styrene was conducted by a 12 member international expert panel selected by the Harvard Centre for Risk Assessment. The scientists had expertise in toxicology, epidemiology, medicine, risk analysis, pharmacokinetics, and exposure assessment.

The Harvard study reported that styrene is naturally present in foods such as strawberries, beef, and spices, and is naturally produced in the processing of foods such as wine and cheese. The study also reviewed all the published data on the quantity of styrene contributing to the diet due to migration of food packaging and disposable food contact articles, and concluded there is no cause for concern for the general public from exposure to styrene from foods or styrenic materials used in food-contact applications, such as polystyrene packaging and food service containers.

Styrene oligomers in polystyrene containers used for food packaging have been found to migrate into the food. Another Japanese study conducted on wild-type and AhR-null mice found that the styrene trimer, which the authors detected in cooked polystyrene container-packed instant foods, may increase thyroid hormone levels.

Fire Hazards

Like other organic compounds, polystyrene is flammable. Polystyrene is classified according to DIN4102 as a "B3" product, meaning highly flammable or "easily ignited." Consequently, although it is an efficient insulator at low temperatures, its use is prohibited in any exposed installations in building construction if the material is not flame retardant. It must be concealed behind drywall, sheet metal or concrete.

Foamed polystyrene plastic materials have been accidentally ignited and caused huge fires and losses, for example at the Düsseldorf International Airport, the Channel tunnel (where polystyrene was inside a railcar that caught fire), and the Browns Ferry Nuclear Power Plant (where fire breached a fire retardant and reached the foamed plastic underneath, inside a firestop that had not been tested and certified in accordance with the final installation).

Cumene

Cumene is the common name for isopropylbenzene, an organic compound that is an aromatic hydrocarbon. It is a constituent of crude oil and refined fuels. It is a flammable colorless liquid that has a boiling point of 152 °C.

Nearly all the cumene that is produced as a pure compound on an industrial scale is converted to cumene hydroperoxide, which is an intermediate in the synthesis of other industrially important chemicals, primarily phenol and acetone.

Production

Commercial production of cumene is carried out through the catalytic alkylation of benzene, with the addition of propylene. Solid phosphoric acid (SPA) supported on alumina can also be used as a catalyst, and this was the case prior to the mid-1990s when zeolite-based catalysts made the other technique commercially redundant. This process is other wise known as Friedel-Crafts alkylation reaction.

Isopropylbenzene is stable, but may form peroxides in storage if in contact with the air. It is important to test for the presence of peroxides before heating or distilling.

The chemical is also flammable and incompatible with strong oxidizing agents. Environmental laboratories commonly test isopropylbenzene using a GCMS instrument.

Cumene Process

The Cumene process (Cumene-phenol process, Hock process) is an industrial process for developing phenol and acetone from benzene and propylene. The term stems from cumene (isopropyl benzene), the intermediate material during the process. It was invented by Heinrich Hock in 1944 and independently by R.

Ûdris and P. Sergeyev in 1942 (USSR). This process converts two relatively cheap starting materials, benzene and propylene, into two more valuable ones, phenol and acetone. Other reactants required are oxygen from air and small amounts of a radical initiator.

Most of the worldwide production of phenol and acetone is now based on this method. In 2003, nearly 7 billion kg of phenol was produced by the Hock Process.

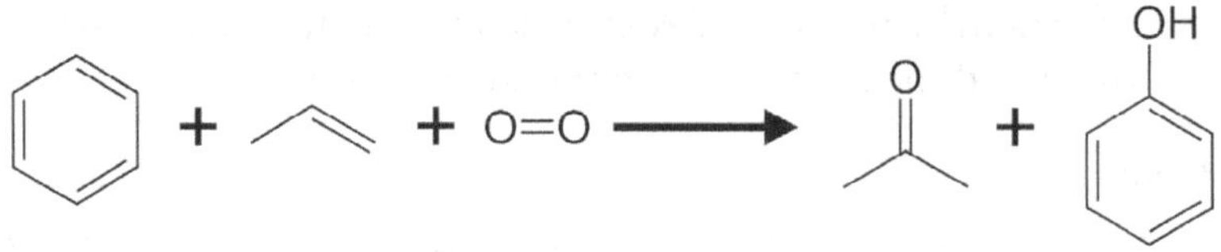

Technical Description

Benzene and propylene are compressed together to a pressure of 30 standard atmospheres at 250 °C (482 °F) in presence of a catalytic Lewis acid. Phosphoric acid is often favoured over aluminium halides. Cumene is formed in the gas-phase Friedel-Crafts alkylation of benzene by propylene:

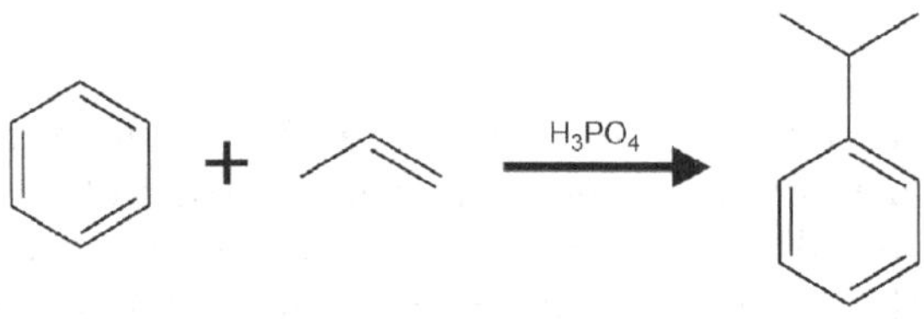

Cumene is oxidized in air which removes the tertiary benzylic hydrogen from cumene and hence forms a cumene radical:

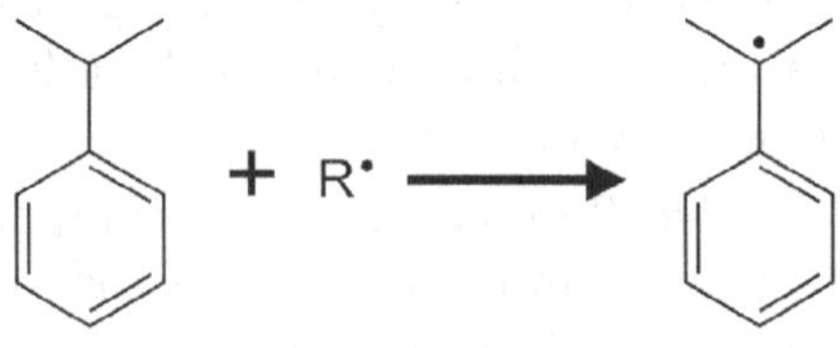

This cumene radical then bonds with an oxygen molecule to give cumene hydroperoxide radical, which in turn forms cumene hydroperoxide ($C_6H_5C(CH_3)_2$-O-O-H) by abstracting benzylic hydrogen from another cumene molecule. This latter cumene converts into cumene radical and feeds back into subsequent chain formations of cumene hydroperoxides. A pressure of 5 atm is used to ensure that the unstable peroxide is kept in liquid state.

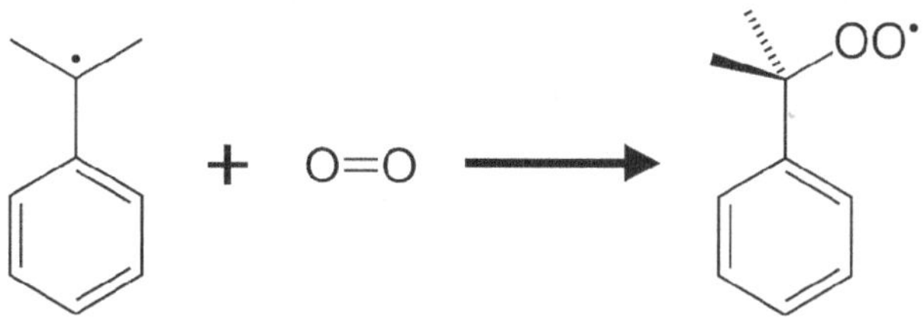

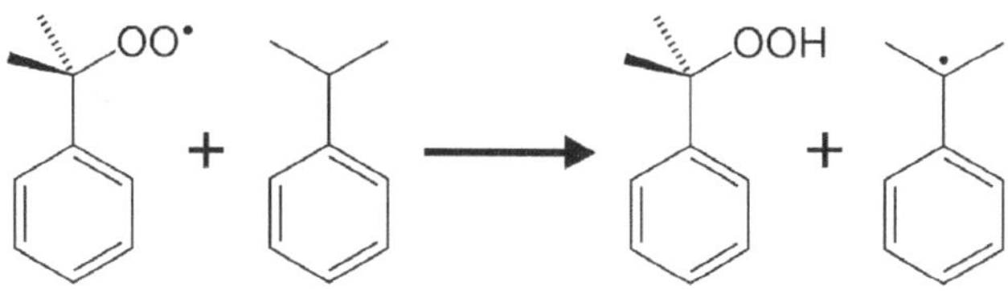

Cumene hydroperoxide is then hydrolysed in an acidic medium (the Hock rearrangement) to give phenol and acetone. In the first step, the terminal hydroperoxy oxygen atom is protonated.

This is followed by a step in which the phenyl group migrates from the benzyl carbon to the adjacent oxygen and a water molecule is lost, producing a resonance stabilized tertiary carbocation.

The concerted mechanism of this step is similar to the mechanisms of the Baeyer-Villiger oxidation and also the oxidation step of hydroboration-oxidation. In 2009, an acidified bentonite clay was proven to be a more economical catalyst than sulfuric acid as the acid medium.

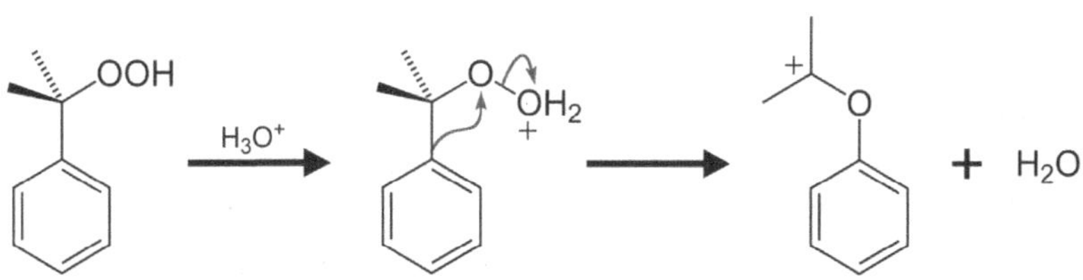

As shown below, the resulting carbocation is then attacked by water, a proton is then transferred from the hydroxy oxygen to the ether oxygen, and finally the ion falls apart into phenol and acetone.

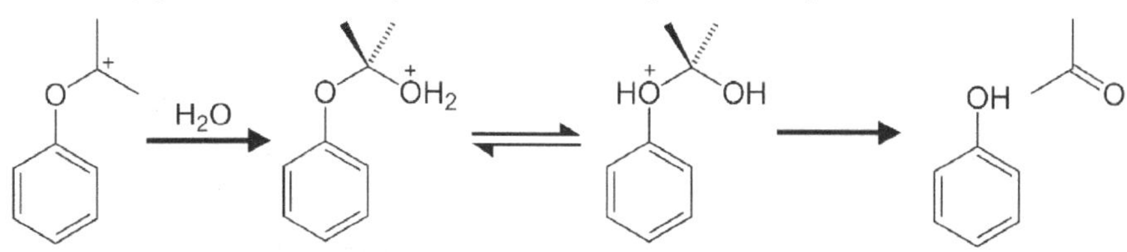

The products are extracted by distillation.

Phenol

Phenol, also known as carbolic acid, is an organic compound with the chemical formula C_6H_5OH. It is a white crystalline solid. The molecule consists of a phenyl (-C_6H_5), bonded to a hydroxyl (-OH) group. It is produced on a large scale (about 7 billion kg/year) as a precursor to many materials and useful compounds. It is only mildly acidic but requires careful handling due to its propensity to cause burns.

Phenol was first extracted from coal tar, and its major uses involve its conversion to plastics or related materials. Phenols are key for building polycarbonates, epoxies, Bakelite, nylon, detergents and a large collection of drugs, herbicides and pharmaceuticals.

Properties

Phenol is appreciably soluble in water, with about 8.3 g dissolving in 100 mL (0.88 M). The sodium salt of phenol, sodium phenoxide, is far more water soluble.

Acidity

It is slightly acidic: the phenol molecules have weak tendencies to lose the H^+ ion from the hydroxyl group, resulting in the highly water-soluble phenolate anion $C_6H_5O^-$ (also called phenoxide). Compared to aliphatic alcohols, phenol is about 1 million times more acidic, although it is still considered a weak acid. It reacts completely with aqueous NaOH to lose H^+, whereas most alcohols react only partially. Phenols are less acidic than carboxylic acids, and even carbonic acid.

One explanation for the increased acidity over alcohols is resonance stabilization of the phenoxide anion by the aromatic ring. In this way, the negative charge on oxygen is shared by the ortho and para carbon atoms. In another explanation, increased acidity is the result of orbital overlap between the oxygen's lone pairs and the aromatic system.

In a third, the dominant effect is the induction from the sp^2 hybridised carbons; the comparatively more powerful inductive withdrawal of electron density that is provided by the sp^2 system compared to an sp^3 system allows for great stabilization of the oxyanion.

In making this conclusion, one can examine the pK_a of the enol of acetone, which is 19.0, in comparison to phenol with a pK_a of 10.0. However, this similarity of acidities of phenol and acetone enol is not observed in the gas phase, and is because the difference of solvation energies of the deprotonated acetone enol and phenoxide almost exactly offsets the experimentally observed gas phase acidity difference. It has recently been shown that only about 1/3 of the increased acidity of phenol is due to inductive effects, with resonance accounting for the rest.

Phenoxide Anion

Phenol can be deprotonated with moderate base such as triethylamine, forming the nucleophilic *phenoxide anion* or *phenolate anion*, which is highly water-soluble.

The phenoxide anion has a similar nucleophilicity to free amines, with the further advantage that its conjugate acid (neutral phenol) does not become entirely deactivated as a nucleophile even in moderately acidic conditions. Phenols are sometimes used in peptide synthesis to "activate" carboxylic acids or esters to form activated esters. Phenolate esters are far more stable than acid anhydrides or acyl halides but are sufficiently reactive under mild conditions to facilitate the formation of amide bonds.

Phenoxides are enolates stabilised by aromaticity. Under normal circumstances, phenoxide is more reactive at the oxygen position, but the oxygen position is a "hard" nucleophile whereas the alpha-carbon positions tend to be "soft".

Tautomerism

Phenol exhibits keto-enol tautomerism with its unstable keto tautomer cyclohexadienone, but only a tiny fraction of phenol exists as the keto form. The equilibrium constant for enolisation is approximately 10, meaning that only one in every ten trillion molecules is in the keto form at any moment. The small amount of stabilisation gained by exchanging a C=C bond for a C=O bond is more than offset by the large destabilisation resulting from the loss of aromaticity. Phenol therefore exists entirely in the enol form.

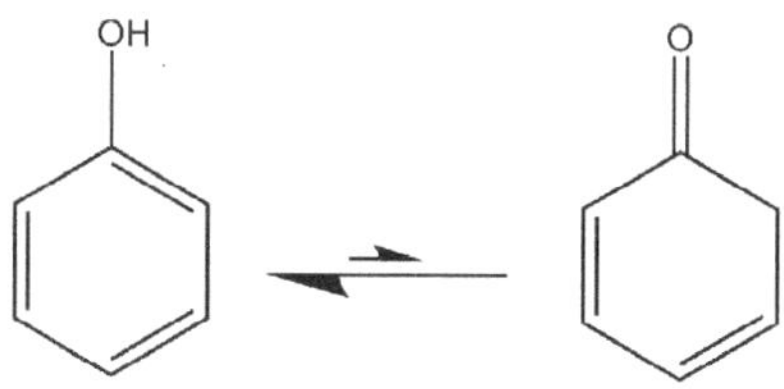

Phenol-cyclohexadienone Tautomerism

Reactions

Phenol is highly reactive toward electrophilic aromatic substitution as the oxygen atom's pi electrons donate electron density into the ring.

By this general approach, many groups can be appended to the ring, via halogenation, acylation, sulfonation, and other processes. However, phenol's ring is so strongly activated - second only to aniline - that bromination or chlorination of phenol leads to substitution on all carbons ortho and para to the hydroxy group, not only on one carbon.

Production

Because of phenol's commercial importance, many methods have been developed for its production. The main route, accounting for 95% of production (2003), involves the partial oxidation of cumene (isopropylbenzene) via the Hock rearrangement:

$$C_6H_5CH(CH_3)_2 + O_2 \rightarrow C_6H_5OH + (CH_3)_2CO$$

An early commercial route begins with the reaction of strong base with benzenesulfonate:

$$C_6H_5SO_3H + 2\ NaOH \rightarrow C_6H_5OH + Na_2SO_3 + H_2O$$

Other methods under consideration involve:

- hydrolysis of chlorobenzene, using base or steam (Raschig-Hooker process):

$$C_6H_5Cl + H_2O \rightarrow C_6H_5OH + HCl$$

- direct oxidation of benzene with nitrous oxide, a potentially "green" process:

$$C_6H_6 + N_2O \rightarrow C_6H_5OH + N_2$$

- oxidation of toluene, as developed by Dow Chemical:

$$C_6H_5CH_3 + 2\ O_2 \rightarrow C_6H_5OH + CO_2 + H_2O$$

In the Lummus Process, the oxidation of toluene to benzoic acid is conducted separately.

Phenol is also recoverable byproduct of coal pyrolysis.

Uses

The major uses of phenol, consuming two thirds of its production, involve its conversion to plastics or related materials. Condensation with acetone gives bisphenol-A, a key precursor to polycarbonates and epoxide resins. Condensation with formaldehyde gives phenolic resins, a famous example of which is Bakelite. Hydrogenation of phenol gives cyclohexanone, a precursor to nylon. Nonionic detergents are produced by alkylation of phenol to give the alkylphenols, e.g., nonylphenol, which are then subjected to ethoxylation.

Phenol is also a versatile precursor to a large collection of drugs, most notably aspirin but also many herbicides and pharmaceuticals.

Phenol is also used as an oral anesthetic/analgesic, commonly used to temporarily treat pharyngitis.

Niche Uses

Phenol is so inexpensive that it attracts many small-scale uses. It once was widely used as an antiseptic, especially as Carbolic soap, from the early 1900s through the 1970s. It is a component of industrial paint strippers used in the aviation industry for the removal of epoxy, polyurethane and other chemically resistant coatings.

Phenol derivatives are also used in the preparation of cosmetics including sunscreens, hair dyes, and skin lightening preparations.

History

Phenol was discovered in 1834, when it was extracted from coal tar, which remained the primary source until the development of the petrochemical industry.

The antiseptic properties of phenol were used by Sir Joseph Lister (1827–1912) in his pioneering technique of antiseptic surgery, although the skin irritation caused by continual exposure to phenol eventually led to the substitution of aseptic (germ-free) techniques in surgery. Lister decided that the wounds themselves had to be thoroughly cleaned. He then covered the wounds with a piece of rag or lint covered in phenol, or carbolic acid as he called it.

It is also the active ingredient in some oral analgesics such as Chloraseptic spray as well as Carmex. Phenol was also the main ingredient of the Carbolic Smoke Ball, an ineffective device marketed in London in the 19th century as protecting the user against influenza and other ailments, and the subject of a famous law case.

Second World War

Injections of phenol have occasionally been used as a means of execution. In particular, phenol and cyanide injections were used as a means of individual execution by the Nazis during the Second World War. Originally used by the Nazis in 1939 as part of Action T4, phenol, inexpensive, easy to make and quickly lethal, became the injectable toxin of choice as part of Nazi Germany's "euthanasia" program. Although Zyklon-B pellets, invented by Gerhard Lenz, were used in the gas chambers to exterminate large groups of people, the Nazis learned that extermination of smaller groups was more economical via injection of each victim, one at a time, with phenol. Phenol injections were given to thousands of people in concentration camps, especially at Auschwitz-Birkenau. Approximately one gram is enough to cause death. Injections were administered by medical doctors, their assistants, or sometimes prisoner doctors; such injections were originally given intravenously, more commonly in the arm, but injection directly into the heart, so as to induce nearly instant death, was later adopted.

One of the best known inmates to be executed with a phenol injection in Auschwitz was St. Maximilian Kolbe, a Catholic priest who volunteered to undergo three weeks of starvation and dehydration in the place of another inmate.

Occurrence

Phenol is a measurable component in the aroma and taste of the distinctive Islay scotch whisky, generally ~30, but up to 100 ppm.

Toxicity

Phenol and its vapours are corrosive to the eyes, the skin, and the respiratory tract. Repeated or prolonged skin contact with phenol may cause dermatitis, or even second and third-degree burns due to phenol's caustic and defatting properties. Inhalation of phenol vapour may cause lung edema.

The substance may cause harmful effects on the central nervous system and heart, resulting in dysrhythmia, seizures, and coma. The kidneys may be affected as well. Exposure may result in death and the effects may be delayed. Long-term or repeated exposure of the substance may have harmful effects on the liver and kidneys."

There is no evidence to believe that phenol causes cancer in humans. Besides its hydrophobic effects, another mechanism for the toxicity of phenol may be the formation of phenoxyl radicals.

Chemical burns from skin exposures can be decontaminated by washing with polyethylene glycol, isopropyl alcohol, or perhaps even copious amounts of water. Removal of contaminated clothing is required, as well as immediate hospital treatment for large splashes. This is particularly important if the phenol is mixed with chloroform (a commonly-used mixture in molecular biology for DNA & RNA purification from proteins).

Phenols

The word *phenol* is also used to refer to any compound that contains a six-membered aromatic ring, bonded directly to a hydroxyl group (-OH). Thus, phenols are a class of organic compounds of which the phenol discussed in this article is the simplest member.

Bisphenol A

Bisphenol A (BPA) is an organic compound with two phenol functional groups. It is used to make polycarbonate plastic and epoxy resins, along with other applications.

As it has been known to be estrogenic since the mid 1930s, concerns about the use of bisphenol A in consumer products were regularly reported in the news media in 2008, after several governments issued reports questioning its safety, prompting some retailers to remove products containing it from their shelves.

A 2010 report from the United States Food and Drug Administration (FDA) raised further concerns regarding exposure of fetuses, infants and young children. In September 2010, Canada became the first country to declare BPA as a toxic substance. In the European Union and Canada, BPA use is banned in baby bottles.

Production

World production capacity of this compound was 1 million tons in the 1980s, and more than 2.2 million tons in 2009. In 2003, U.S. consumption was 856,000 tons, 72% of which was used to make polycarbonate plastic and 21% going into epoxy resins. In the US, less than 5% of the BPA produced is used in food contact applications.

Bisphenol A was first synthesized by the Russian chemist A.P. Dianin in 1891. This compound is synthesized by the condensation of acetone (hence the suffix A in the name) with two equivalents of phenol. The reaction is catalysed by a strong acid, such as hydrochloric acid (HCl) or a sulfonated polystyrene resin.

Industrially, a large excess of phenol is used to ensure full condensation; the product mixture of the cumene process (acetone and phenol) may also be used as starting material:

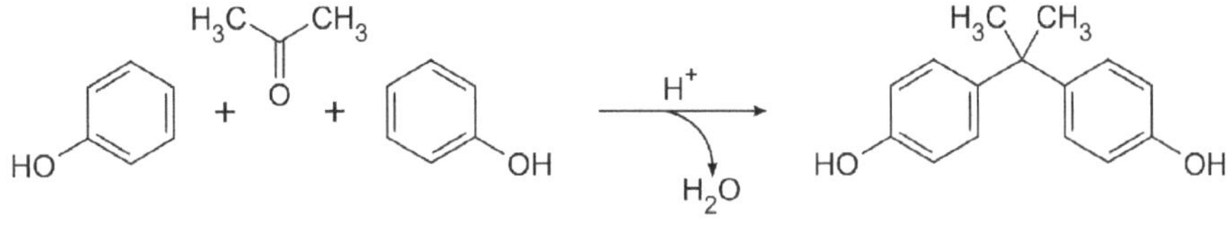

A large number of ketones undergo analogous condensation reactions. Commercial production of BPA requires distillation – either extraction of BPA from many resinous byproducts under high vacuum, or solvent-based extraction using additional phenol followed by distillation.

Use

Bisphenol A is used primarily to make plastics, and products containing bisphenol A-based plastics have been in commerce use since 1957. At least 8 billion pounds of BPA are used by manufacturers yearly. It is a key monomer in production of epoxy resins and in the most common form of polycarbonate plastic. The overall reaction to give polycarbonate can be written:

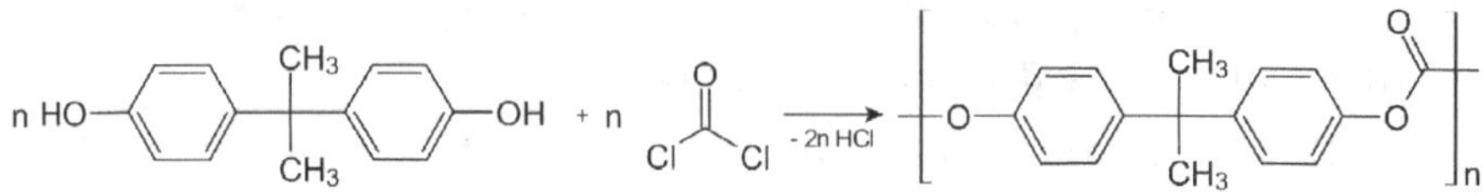

Polycarbonate plastic, which is clear and nearly shatter-proof, is used to make a variety of common products including baby and water bottles, sports equipment, medical and dental devices, dental fillings and sealants, eyeglass lenses, CDs and DVDs, and household electronics. BPA is also used in the synthesis of polysulfones and polyether ketones, as an antioxidant in some plasticizers, and as a polymerization inhibitor in PVC. Epoxy resins containing bisphenol A are used as coatings on the inside of almost all food and beverage cans, however, due to BPA health concerns, in Japan epoxy coating was mostly replaced by PET film. Bisphenol A is also a precursor to the flame retardant tetrabromobisphenol A, and was formerly used as a fungicide.

Bisphenol A is a preferred colour developer in carbonless copy paper and thermal paper, with the most common public exposure coming from some thermal point of sale receipt paper. BPA-based products are also used in foundry castings and for lining water pipes.

Identification in Plastics

"In general, plastics that are marked with recycle codes 1, 2, 4, 5, and 6 are very unlikely to contain BPA. Some, but not all, plastics that are marked with recycle codes 3 or 7 may be made with BPA."

There are seven classes of plastics used in packaging applications. Type 7 is the catch-all "other" class, and some type 7 plastics, such as polycarbonate (sometimes identified with the letters "PC" near the recycling symbol) and epoxy resins, are made from bisphenol A monomer.

Type 3 (PVC) can also contain bisphenol A as an antioxidant in plasticizers. This is particularly true for "flexible PVC", but not true for PVC pipes.

Health Effects

Bisphenol A is an endocrine disruptor, which can mimic the body's own hormones and may lead to negative health effects. Early development appears to be the period of greatest sensitivity to its effects, and some studies have linked prenatal exposure to later neurological difficulties.

Regulatory bodies have determined safety levels for humans, but those safety levels are currently being questioned or under review as a result of new scientific studies. A 2011 study that investigated the number of chemicals pregnant women are exposed to in the U.S. found BPA in 96% of women.

In 2009, The Endocrine Society released a statement expressing concern over current human exposure to BPA.

In 2011, the Food Standards Agency's chief scientist said "the evidence [is] that BPA is rapidly absorbed, detoxified, and eliminated from humans – therefore is not a health concern."

Expert Panel Conclusions

In 2007, a consensus statement by 38 experts on bisphenol A concluded that average levels in people are above those that cause harm to many animals in laboratory experiments. However, they noted that while BPA is not persistent in the environment or in humans, biomonitoring surveys indicate that exposure is continuous, which is problematic because acute animal exposure studies are used to estimate daily human exposure to BPA, and no studies that had examined BPA pharmacokinetics in animal models had followed continuous low level exposures.

They added that measurement of BPA levels in serum and other body fluids suggests that either BPA intake is much higher than accounted for, or that BPA can bioaccumulate in some conditions such as pregnancy, or both. A 2011 study, the first to examine BPA in a continuous low level exposure throughout the day, did find an increased absorption and accumulation of BPA in the blood of mice.

In 2007 it was reported that among government-funded BPA experiments on lab animals and tissues, 153 found adverse effects and 14 did not, whereas all 13 studies funded by chemical corporations reported no harm. The studies indicating harm reported a variety of deleterious effects in rodent offspring exposed in the womb: abnormal weight gain, insulin resistance, prostate cancer, and excessive mammary gland development.

A panel convened by the U.S. National Institutes of Health in 2007 determined that there was "some concern" about BPA's effects on fetal and infant brain development and behaviour. The concern over the effect of BPA on infants was also heightened by the fact that infants and children are estimated to have the highest daily intake of BPA.

A 2008 report by the U.S. National Toxicology Program (NTP) later agreed with the panel, expressing "some concern for effects on the brain, behaviour, and prostate gland in fetuses, infants, and children at current human exposures to bisphenol A," and "*minimal* concern for effects on the mammary gland and an earlier age for puberty for females in fetuses, infants, and children at current human exposures to bisphenol A."

The NTP had "*negligible* concern that exposure of pregnant women to bisphenol A will result in fetal or neonatal mortality, birth defects, or reduced birth weight and growth in their offspring."

Obesity

A 2008 review has concluded that obesity may be increased as a function of BPA exposure, which "...merits concern among scientists and public health officials." A 2009 review of available studies has concluded that "perinatal BPA exposure acts to exert persistent effects on body weight and adiposity". Another 2009 review has concluded that "Eliminating exposures to (BPA) and improving nutrition during development offer the potential for reducing obesity and associated diseases".

Other reviews have come with similar conclusions. A later study on rats has suggested that perinatal exposure to drinking water containing 1 mg/L of BPA increased adipogenesis in females at weaning. Other study suggested that larger size-for-age was due to a faster growth rate rather than obesity.

Neurological Issues

A panel convened by the U.S. National Institutes of Health determined that there was "some concern" about BPA's effects on fetal and infant brain development and behaviour. A 2008 report by the U.S. National Toxicology Program (NTP) later agreed with the panel, expressing "some concern for effects on the brain". In January 2010 the FDA expressed the same level of concern.

A 2007 review has concluded that BPA, like other xenoestrogens, should be considered as a player within the nervous system that can regulate or alter its functions through multiple pathways. A 2007 review has concluded that low doses of BPA during development have persistent effects on brain structure, function and behaviour in rats and mice.

A 2008 review concluded that low-dose BPA maternal exposure causes long-term consequences at the level of neurobehavioural development in mice. A 2008 review has concluded that neonatal exposure to Bisphenol-A (BPA) can affect sexually dimorphic brain morphology and neuronal adult phenotypes in mice. A 2008 review has concluded that BPA altered long-term potentiation in the hippocampus and even nanomolar dosage could induce significant effects on memory processes. A 2009 review raised concerns about BPA effect on anteroventral periventricular nucleus.

A 2008 study by the Yale School of Medicine demonstrated that adverse neurological effects occur in non-human primates regularly exposed to bisphenol A at levels equal to the United States Environmental Protection Agency's (EPA) maximum safe dose of 50 μg/kg/day. This research found a connection between BPA and interference with brain cell connections vital to memory, learning and mood.

A 2010 study with rats prenatally exposed to 40 microg/kg bw BPA has concluded that corticosterone and its actions in the brain are sensitive to the programming effects of BPA.

Disruption of the Dopaminergic System

A 2005 review concluded that prenatal and neonatal exposure to BPA in mice can potentiate the central dopaminergic systems, resulting in the supersensitivity to the drugs-of-abuse-induced reward effects and hyperlocomotion.

A 2008 review has concluded that BPA mimics estrogenic activity and impacts various dopaminergic processes to enhance mesolimbic dopamine activity resulting in hyperactivity, attention deficits, and a heightened sensitivity to drugs of abuse.

A 2009 study on rats has concluded that prenatal and neonatal exposure to low-dose BPA causes deficits in development at dorsolateral striatum via altering the function of dopaminergic receptors. Another 2009 study has found associated changes in the dopaminergic system.

Thyroid Function

A 2007 review has concluded that bisphenol-A has been shown to bind to thyroid hormone receptor and perhaps have selective effects on its functions.

A 2009 review about environmental chemicals and thyroid function, raised concerns about BPA effects on triiodothyronine and concluded that "available evidence suggests that governing agencies need to regulate the use of thyroid-disrupting chemicals, particularly as such uses relate exposures of pregnant women, neonates and small children to the agents".

A 2009 review summarized BPA adverse effects on thyroid hormone action.

Cancer Research

According to the WHO's INFOSAN, "animal studies have not provided convincing evidence of risk of cancer from BPA exposure."

Neither the U.S. Environmental Protection Agency nor the International Agency for Research on Cancer has evaluated bisphenol A for possible carcinogenic activity.

A 2010 review at Tufts University Medical School concluded that Bisphenol A may increase cancer risk.

Breast Cancer

A 2008 review has concluded that "perinatal exposure to (...) low doses of (..) BPA, alters breast development and increases breast cancer risk". Another 2008 review concluded that "animal experiments and epidemiological data strengthen the hypothesis that fetal exposure to xenoestrogens may be an underlying cause of the increased incidence of breast cancer observed over the last 50 years".

A 2009 in vitro study has concluded that BPA is able to induce neoplastic transformation in human breast epithelial cells. Another 2009 study concluded that maternal oral exposure to low concentrations of BPA during lactation increases mammary carcinogenesis in a rodent model.

A 2010 study with the mammary glands of the offspring of pregnant rats treated orally with 0, 25 or 250 μg BPA/kg body weight has found that key proteins involved in signaling pathways such as cellular proliferation were regulated at the protein level by BPA.

A 2010 study has found that BPA may reduce sensitivity to chemotherapy treatment of specific tumours.

Neuroblastoma

In vitro studies have suggested that BPA can promote the growth of neuroblastoma cells. A 2010 in vitro study has concluded that BPA potently promote invasion and metastasis of neuroblastoma cells through overexpression of MMP-2 and MMP-9 as well as downregulation of TIMP2.

Prostate Development and Cancer

A 1997 study in mice has found that neonatal BPA exposure of 2 μg/kg increased adult prostate weight. A 2005 study in mice has found that neonatal BPA exposure at 10 μg/kg disrupted the development of the fetal mouse prostate.

A 2006 study in rats has shown that neonatal bisphenol A exposure at 10 μg/kg levels increases prostate gland susceptibility to adult-onset precancerous lesions and hormonal carcinogenesis. A 2007 in vitro study has found that BPA within the range of concentrations currently measured in human serum is associated with permanent increases in prostate size. A 2009 study has found that newborn rats exposed to a low-dose of BPA (10 μg/kg) increased prostate cancer susceptibility when adults.

DNA Methylation

At least one study has suggested that bisphenol A suppresses DNA methylation which is linked to epigenetic changes.

Reproductive System and Sexual Behaviour Research

A 2007 study using pregnant mice showed that BPA changes the expression of key developmental genes that form the uterus, which may impact female reproductive tract development and future fertility of female fetuses.

A series of studies made in 2009 found:

- Mouse ovary anomalies from exposure as low as 1 μg/kg, concluded that BPA exposure causes long-term adverse reproductive and carcinogenic effects if exposure occurs during prenatal critical periods of differentiation.
- Neonatal exposure of as low as 50 μg/kg disrupts ovarian development in mice.
- Neonatal BPA exposition of as low as 50 μg/kg permanently alters the hypothalamic estrogen-dependent mechanisms that govern sexual behaviour in the adult female rat.
- Prenatal exposure to BPA at levels of (10 μg/kg/day) affects behavioural sexual differentiation in male monkeys.
- In placental JEG3 cells *in vitro* BPA may reduce estrogen synthesis.

- BPA exposure disrupted the blood-testis barrier when administered to immature, but not to adult, rats.
- Exposure to BPA in the workplace was associated with self-reported adult male sexual dysfunction.

A 2009 rodent study, funded by EPA and conducted by some of its scientists, concluded that, compared with ethinyl estradiol, low-dose exposures of bisphenol A (BPA) showed no effects on several reproductive functions and behavioural activities measured in female rats. That study was criticized as flawed for using polycarbonate cages in the experiment (since polycarbonate contains BPA) and the claimed resistance of the rats to estradiol, but that claim was contested by the authors and others.

Another 2009 rodent study found that BPA exposure during pregnancy has a lasting effect on one of the genes that is responsible for uterine development and subsequent fertility in both mice and humans (HOXA10). The authors concluded, "We don't know what a safe level of BPA is, so pregnant women should avoid BPA exposure."

In a 2010 study mice were given BPA at doses thought to be equivalent to levels currently being experienced by humans. The research showed that BPA exposure affects the earliest stages of egg production in the ovaries of the developing mouse fetuses, thus suggesting that the next generation may suffer genetic defects in such biological processes as mitosis and DNA replication.

In addition, the research team noted that their study "revealed a striking down-regulation of mitotic/cell cycle genes, raising the possibility that BPA exposure immediately before meiotic entry might act to shorten the reproductive lifespan of the female" by reducing the total pool of fetal oocytes.

Another 2010 study with mice concluded that BPA exposure in utero leads to permanent DNA alterations in sensitivity to estrogen. Also in 2010, a rodent study found that by exposing fetal mice to BPA during pregnancy and examining gene expression and DNA in the uteruses of female fetuses, BPA exposure permanently affected the uterus by decreasing regulation of gene expression.

The changes caused the mice to over-respond to estrogen throughout adulthood, long after the BPA exposure, thus suggesting that early exposure to BPA genetically "programmed" the uterus to be hyper-responsive to estrogen. Extreme estrogen sensitivity can lead to fertility problems, advanced puberty, altered mammary development and reproductive function, as well as a variety of hormone-related cancers. One of the authors concluded that BPA may be similar to diethylstilbestrol caused birth defects and cancers in young women whose mothers were given the drug during pregnancy.

A 2011 study using the rhesus monkey, a species that is very similar to humans in regard to pregnancy and fetal development, found that prenatal exposure to BPA causes changes in female primates' uterus development. A 2011 rodent study found that male rats exposed to BPA had lower sperm counts and testosterone levels than those of unexposed males. A 2011 mice study found that male mice exposed to BPA became demasculinized and behaved more like females in their spatial navigational abilities. They were also less desirable to female mice.

General Research

At an Endocrine Society meeting in 2009, new research reported data from animals experimentally treated with BPA. Studies presented at the group's annual meeting show BPA can affect the hearts of women, can permanently damage the DNA of mice, and appear to be entering the human body from a variety of unknown sources.

A 2009 *in vitro* study on cytotrophoblasts cells has found cytoxic effects in exposure of BPA doses from 0.0002 to 0.2 micrograms per millilitre and concluded this finding "suggests that exposure of placental cells to low doses of BPA may cause detrimental effects, leading *in vivo* to adverse pregnancy outcomes such as preeclampsia, intrauterine growth restriction, prematurity and pregnancy loss"

A 2009 study in rats concluded that BPA, at the reference safe limit for human exposure, was found to impact intestinal permeability and may represent a risk factor in female offspring for developing severe colonic inflammation in adulthood.

A 2010 study on mice has concluded that perinatal exposure to 10 micrograms/mL of BPA in drinking water enhances allergic sensitization and bronchial inflammation and responsiveness in an animal model of asthma, and a 2011 study found that higher BPA concentrations in the urine of the pregnant women at 16 weeks were associated with wheezing, a symptom of asthma, in their babies.

Studies on Humans

Lang Study and Heart Disease

The first large study of health effects on humans associated with bisphenol A exposure was published in September 2008 by Iain Lang and colleagues in the *Journal of the American Medical Association.* The cross-sectional study of almost 1,500 people assessed exposure to bisphenol A by looking at levels of the chemical in urine. The authors found that higher bisphenol A levels were significantly associated with heart disease, diabetes, and abnormally high levels of certain liver enzymes. An editorial in the same issue concludes:

"Based on this background information, the study by Lang et al, while preliminary with regard to these diseases in humans, should spur US regulatory agencies to follow the recent action taken by Canadian regulatory agencies, which have declared BPA a "toxic chemical" requiring aggressive action to limit human and environmental exposures.

Alternatively, Congressional action could follow the precedent set with the recent passage of federal legislation designed to limit exposures to another family of compounds, phthalates, also used in plastic. Like BPA, phthalates are detectable in virtually everyone in the United States. This bill moves US policy closer to the European model, in which industry must provide data on the safety of a chemical before it can be used in products."

A later similar study performed by the same group of scientists, published in January 2010, confirmed, despite of lower concentrations of BPA in the second study sample, an associated increased risk for heart disease but not for diabetes or liver enzymes. Patients with the highest levels of BPA in their urine carried a 33% increased risk of coronary heart disease.

Other Studies

Studies have associated recurrent miscarriage with BPA serum concentrations, oxidative stress and inflammation in postmenopausal women with urinary concentrations, externalizing behaviours in two-year old children, especially among female children, with mother's urinary concentrations, altered hormone levels in men and declining male sexual function with urinary concentrations.

The Canadian Health Measures Survey, 2007 to 2009 published in 2010 found that teenagers carry 30 percent more l bisphenol A (BPA) in their bodies than older adults. The reason for this is not known. A 2010 study that analyzed BPA urinary concentrations has concluded that for people under 18 years of age BPA may negatively impact human immune function.

A study done in 2010 reported the daily excretion levels of BPA among European adults in a large-scale and high-quality population-based sample, and it was shown that higher BPA daily excretion was associated with an increase in serum total testosterone concentration in men. A 2011 study found higher BPA levels in women with polycystic ovary syndrome compared to controls.

Furthermore, researchers found a statistically significant positive association between male sex hormones and BPA in these women suggesting a potential role of BPA in ovarian dysfunction. A 2010 study found that people over age 18 with higher levels of BPA exposure had higher CMV antibody levels, which suggests their cell-mediated immune system may not be functioning properly.

Sexual Difficulties

A 2009 study on Chinese workers in BPA factories found that workers were four times more likely to report erectile dysfunction, reduced sexual desire and overall dissatisfaction with their sex life than workers with no heightened BPA exposure. BPA workers were also seven times more likely to have ejaculation difficulties. They were also more likely to report reduced sexual function within one year of beginning employment at the factory, and the higher the exposure, the more likely they were to have sexual difficulties.

Historical Studies

The first evidence of the estrogenicity of bisphenol A came from experiments on rats conducted in the 1930s, but it was not until 1997 that adverse effects of low-dose exposure on laboratory animals were first reported.

Low-dose Exposure in Animals

Dose (μg/kg/day)	*Effects (measured in studies of mice or rats,descriptions (in quotes) are from Environmental Working Group)*	*Study Year*
0.025	"Permanent changes to genital tract"	2005
0.025	"Changes in breast tissue that predispose cells to hormones and carcinogens"	2005
1	long-term adverse reproductive and carcinogenic effects	2009
2	"increased prostate weight 30%"	1997
2	"lower bodyweight, increase of anogenital distance in both genders, signs of early puberty and longer estrus."	2002
2.4	"Decline in testicular testosterone"	2004

Contd...

Dose (μg/kg/day)	**Effects (measured in studies of mice or rats,descriptions (in quotes) are from Environmental Working Group)**	**Study Year**
2.5	"Breast cells predisposed to cancer"	2007
10	"Prostate cells more sensitive to hormones and cancer"	2006
10	"Decreased maternal behaviours"	2002
30	"Reversed the normal sex differences in brain structure and behaviour"	2003
50	Adverse neurological effects occur in non-human primates	2008
50	Disrupts ovarian development	2009

The current U.S. human exposure limit set by the EPA is 50 μg/kg/day.

Xenoestrogen

There is evidence that bisphenol A functions as a xenoestrogen by binding strongly to estrogen-related receptor γ (ERR-γ). This orphan receptor (endogenous ligand unknown) behaves as a constitutive activator of transcription. BPA seems to bind strongly to ERR-γ (dissociation constant = 5.5 nM), but not to the estrogen receptor (ER). BPA binding to ERR-γ preserves its basal constitutive activity. It can also protect it from deactivation from the selective estrogen receptor modulator 4-hydroxytamoxifen.

Different expression of ERR-γ in different parts of the body may account for variations in bisphenol A effects. For instance, ERR-γ has been found in high concentration in the placenta, explaining reports of high bisphenol accumulation in this tissue.

Human Exposure Sources

"The problem is, BPA is also a synthetic estrogen, and plastics with BPA can break down, especially when they're washed, heated or stressed, allowing the chemical to leach into food and water and then enter the human body. That happens to nearly all of us; the CDC has found BPA in the urine of 93% of surveyed Americans over the age of 6. If you don't have BPA in your body, you're not living in the modern world." —The Perils of Plastic, TIME Magazine.

Bisphenol A has been known to be leached from the plastic lining of canned foods and polycarbonate plastics, especially those cleaned with harsh detergents or that contain acidic or high-temperature liquids. BPA is an ingredient in the internal coating of metal food and beverage cans used to protect the food from direct contact with the can. A recent Health Canada study found that the majority of canned soft drinks it tested had low, but measurable levels of bisphenol A.

Furthermore, A study conducted by the University of Texas School of Public Health in 2010, found BPA in 63 of 105 samples of fresh and canned foods, foods sold in plastic packaging, and in cat and dog foods in cans and plastic packaging. This included fresh turkey, canned green beans, and canned infant formula. While most human exposure is through diet, exposure can also occur through air and through skin absorption.

A 2011 study published in *Environmental Health Perspectives*, "Food Packaging and Bisphenol A and Bis(2-Ethyhexyl) Phthalate Exposure: Findings from a Dietary Intervention," selected 20 participants based on their self-reported use of canned and packaged foods to study BPA. Participants ate their usual diets, followed by three days of consuming foods that were not canned or packaged.

The study's findings include: 1) evidence of BPA in participants' urine decreased by 50% to 70% during the period of eating fresh foods; and 2), participants' reports of their food practices suggested that consumption of canned foods and beverages and restaurant meals were the most likely sources of exposure to BPA in their usual diets. The researchers note that, even beyond these 20 participants, BPA exposure is widespread, with detectable levels in urine samples in more than an estimated 90% of the U.S. population.

Free BPA is found in high concentration in thermal paper and carbonless copy paper, which would be expected to be more available for exposure than BPA bound into resin or plastic. Popular uses of thermal paper include receipts, event and cinema tickets, labels, and airline tickets.

A Swiss study found that 11 of 13 thermal printing papers contained 8 – 17 g/kg Bisphenol A (BPA). Upon dry finger contact with a thermal paper receipt, roughly 1 μg BPA (0.2 – 6 μg) was transferred to the forefinger and the middle finger. For wet or greasy fingers approximately 10 times more was transferred. Extraction of BPA from the fingers was possible up to 2 hours after exposure. While there is little concern for dermal absorption of BPA, free BPA can readily be transferred to skin and residues on hands can be ingested. Studies conducted by the CDC found bisphenol A in the urine of 95% of adults sampled in 1988–1994 and in 93% of children and adults tested in

2003–04. While the EPA considers exposures up to 50 μg/kg/day to be safe, the most sensitive animal studies show effects at much lower doses.

In 2009, a study found that drinking from polycarbonate bottles increased urinary bisphenol A levels by two thirds, from 1.2 micrograms/gram creatinine to 2 micrograms/gram creatinine. Consumer groups recommend that people wishing to lower their exposure to bisphenol A avoid canned food and polycarbonate plastic containers (which shares resin identification code 7 with many other plastics) unless the packaging indicates the plastic is bisphenol A-free.

To avoid the possibility of BPA leaching into food or drink, the National Toxicology Panel recommends avoiding microwaving food in plastic containers, putting plastics in the dishwasher, or using harsh detergents. In the U.S., consumption of soda, school lunches, and meals prepared outside the home was statistically significantly associated with higher urinary BPA.

BPA is also used to form epoxy resin coating of water pipes. In older buildings, such resin coatings are used to avoid replacement of deteriorating hot and cold water pipes.

Fetal and Early-childhood Exposures

Children may be more susceptible to BPA exposure than adults. A recent study found higher urinary concentrations in young children than in adults under typical exposure scenarios. This increased susceptibility is most likely based on their reduced capacity to eliminate xenobiotics and also their estimated higher daily exposure to BPA, adjusted for weight, compared to adults.

Infants fed with liquid formula are among the most exposed, and those fed formula from polycarbonate bottles can consume up to 13 micrograms of bisphenol A per kg of body weight per day (μg/kg/day).

In the US and Canada, BPA has been found in infant liquid formula in concentrations varying from 0.48 to 11 ng/g. BPA has been rarely found in infant powder formula (only 1 of 14). While breast milk is the optimal source of nutrition for infants, it is not always an option. The U.S. Department of Health & Human Services (HHS) states that "the benefit of a stable source of good nutrition from infant formula and food outweighs the potential risk of BPA exposure.".

A 2010 study of people in Austria, Switzerland, and Germany has suggested polycarbonate (PC) baby bottles as the most prominent role of exposure for infants, and canned food for adults and teenagers. In the United States, the growing concern over BPA exposure in infants in recent years has lead the manufacturers of plastic baby bottles to stop using BPA in their bottles. However, babies may still be exposed if they are fed with old or hand-me-down bottles bought before the companies stopped using BPA.

One often overlooked source of exposure occurs when a pregnant woman is exposed, thereby exposing the fetus. Animal studies have shown that BPA can be found in both the placenta and the amniotic fluid of pregnant mice.

A small US study in 2009, funded by the EWG, detected an average of 2.8 ng/mL BPA in the blood of 9 out of the 10 umbilical cords tested. After the baby is born, maternal exposure can continue to effect the infant through transfer of BPA to the infant via breast milk. Because of these exposures that can occur both during and after pregnancy, mothers wishing to limit their child's exposure to BPA should attempt to limit their own exposures during that time period.

While the majority of exposures have been shown to come through the diet, accidental ingestion can also be considered a source of exposure. One study conducted in Japan tested plastic baby books to look for possible leaching into saliva when babies chew on them. While the results of this study have yet to be replicated, it gives reason to question whether exposure can also occur in infants through ingestion by chewing on certain books or toys.

Population	***Estimated daily bisphenol A intake, μg/kg/day. Table adapted from the National Toxicology Program Expert Panel Report.***
Infant (0–6 months)formula-fed	1–24
Infant (0–6 months)breast-fed	0.2–1
Infant (6–12 months)	1.65–13
Child (1.5–6 years)	0.043–14.7
Adult	0.008–1.5

Pharmacokinetics

There is no agreement between scientists of a physiologically-based pharmacokinetic (PBPK) BPA model for humans. The effects of BPA on an organism depend on how much free BPA is available and for how long cells are exposed to it. Glucuronidation in the liver, by conjugation with glucuronic acid to form the metabolite BPA-glucuronide (BPAG), reduces the amount of free BPA, however BPAG can be deconjugated by beta-glucuronidase, an enzyme present in high concentration in placenta and other tissues. Free BPA can also be inactivated by sulfation, a process that can also be reverted by arylsulfatase C. A 2010 vitro study has shown that placenta P-glycoprotein may efflux BPA from placenta.

The best test methods for studying BPA effects are currently under discussion with scientists sharing different opinions.

A 2010 review of 80+ biomonitoring studies concluded that the general population is internally exposed to significant amounts of unconjugated BPA (in the ng/ml blood range). Using GC/MS on 20 samples, BPA was detected in 100% of urine samples with a median of 1.25 ng/ml, and 10% of blood samples (LOD 0.5 ng/ml).

A 2009 research has found that some drugs, like naproxen, salicylic acid, carbamazepine and mefenamic acid can, *in vitro*, significantly inhibit BPA glucuronidation.

A 2010 study on rats embryos has found that genistein may enhance developmental toxicity of BPA.

Environmental Risk

In general, studies have shown that BPA can affect growth, reproduction and development in aquatic organisms. Among freshwater organisms, fish appear to be the most sensitive species. Evidence of endocrine-related effects in fish, aquatic invertebrates, amphibians and reptiles has been reported at environmentally relevant exposure levels lower than those required for acute toxicity. There is a widespread variation in reported values for endocrine-related effects, but many fall in the range of 1μg/L to 1 mg/L.

BPA can contaminate the environment either directly or through degradation of products containing BPA, such as ocean-borne plastic trash.

As an environmental contaminant this compound interferes with nitrogen fixation at the roots of leguminous plants associated with the bacterial symbiont *Sinorhizobium meliloti*. Despite a half-life in the soil of only 1–10 days, its ubiquity makes it an important pollutant. According to Environment Canada, "initial assessment shows that at low levels, bisphenol A can harm fish and organisms over time. Studies also indicate that it can currently be found in municipal wastewater."

A 2009 review of the biological impacts of plasticizers on wildlife published by the Royal Society with a focus on annelids (both aquatic and terrestrial), molluscs, crustaceans, insects, fish and amphibians concluded that BPA have been shown to affect reproduction in all studied animal groups, to impair development in crustaceans and amphibians and to induce genetic aberrations.

A large 2010 study of two rivers in Canada found that areas contaminated with hormone-like chemicals including bisphenol A showed females made up 85 per cent of the population of a certain fish, while females made up only 55 per cent in uncontaminated areas.

Government and Industry Response

World Health Organization

Arguing uncertainty of possible adverse health effects of low dose BPA exposure, especially on the nervous system and on behaviour, and also the differences of exposure of very young children, the WHO announced in November 2009 that it would organize an expert consultation in 2010 to assess BPA safety.

The WHO expert panel recommended no new regulations limiting or banning the use of Bisphenol-A; stating that "initiation of public health measures would be premature."

Australia and New Zealand

The Australia and New Zealand Food Safety Authority (Food Standards Australia New Zealand) does not see any health risk with bisphenol A baby bottles if the manufacturer's instructions are followed. Levels of exposure are very low and do not pose a significant health risk. It added that "the move by overseas manufacturers to stop using BPA in baby bottles is a voluntary action and not the result of a specific action by regulators." It suggests the use of glass baby bottles if parents have any concerns.

Canada

In April 2008, Health Canada concluded that, while adverse health effects were not expected, the margin of safety was too small for formula-fed infants and proposed classifying the chemical as "'toxic' to human health and the environment."

After the release of that assessment, Canadian Health Minister Tony Clement announced Canada's intent to ban the import, sale, and advertisement of polycarbonate baby bottles containing bisphenol A due to safety concerns, and investigate ways to reduce BPA contamination of baby formula packaged in metal cans. While the agency concluded that human exposures were less than levels believed unsafe, the margin of safety was not high enough for formula-fed infants. Around the same time, Wal-Mart announced that it was immediately ceasing sales in all its Canadian stores of food containers, water and baby bottles, sippy cups, and pacifiers containing bisphenol A, and that it would phase out baby bottles made with it in U.S. stores by early 2009. Nalgene also announced it will stop using the chemical in its products, and Toys-R-Us said it too will cease selling baby bottles made from it. Subsequent news reports showed many retailers removing polycarbonate drinking products from their shelves.

The federal government proposed declaring Bisphenol A a hazardous substance in October 2008 and has since placed it on its list of toxic substances. Health officials wrote in *Canada Gazette* that "It is concluded that bisphenol A be considered as a substance that may be entering the environment in a quantity or concentration or under conditions that constitute or may constitute a danger in Canada to human life or health."

The federal ministries of health and the environment announced they would seek to restrict imports, sales and advertising of polycarbonate baby bottles containing BPA.

In its statement released on 18 October 2008, Health Canada noted that "bisphenol A exposure to newborns and infants is below levels that cause effects" and that the "general public need not be concerned".

Environment Canada listed bisphenol A as a "toxic substance" in September 2010.

European Union

The updated 2008 European Union Risk Assessment Report on bisphenol A, published in June 2008, by the European Commission and European Food Safety Authority (EFSA), concluded that bisphenol A-based products, such as polycarbonate plastic and epoxy resins, are safe for consumers and the environment when used as intended.

By October 2008, after the Lang Study was published, the EFSA issued a statement concluding that the study provided no grounds to revise the current Tolerable Daily Intake (TDI) level for BPA of 0.05 mg/kg bodyweight.

A 2009 scientific study criticized the European risk assessment processes of endocrine disruptors, including BPA.

On 22 December 2009, the EU Environment ministers released a statement expressing concerns over recent studies showing adverse effects of exposure to endocrine disruptors.

In September 2010, the European Food Safety Authority (EFSA) published its latest scientific opinion, based on a "comprehensive evaluation of recent toxicity data [...] concluded that no new study could be identified, which would call for a revision of the current TDI".

The Panel noted that some studies conducted on developing animals have suggested BPA-related effects of possible toxicological relevance, in particular biochemical changes in brain, immune-modulatory effects and enhanced susceptibility to breast tumours but considered that those studies had several shortcomings so the relevance of these findings for human health could not be assessed.

On 25 November 2010, the European Union executive commission said it will ban the manufacturing by 1 March 2011 and ban the marketing and market placement of polycarbonate baby bottles containing the organic compound bisphenol A (BPA) by 1 June 2011, according to John Dalli, commissioner in charge of health and consumer policy. This is backed by a majority of EU governments.

The ban was called an over-reaction by Richard Sharpe, of the Medical Research Council's Human Reproductive Sciences Unit, who said to be unaware of any convincing evidence justifying the measure and criticized it as being done on political, rather than scientific grounds.

Denmark

In May 2009, the Danish parliament passed a resolution to ban the use of BPA in baby bottles, which has not been enacted by April 2010. In March 2010, a temporary ban was declared by the Health Minister.

Belgium

On March 2010, senator Philippe Mahoux proposed legislation to ban BPA in food contact plastics.

France

On 5 February 2010, the French Food Safety Agency (AFSSA) questioned the previous assessments of the health risks of BPA, especially in regard to behavioural effects observed in rat pups following exposure in utero and during the first months of life. In April 2010, the AFFSA suggested the adoption of better labels for food products containing BPA.

On 24 March 2010, French Senate unanimously approved a proposition of law to ban BPA from baby bottles. The National Assembly (Lower House) approved the text on 23 June 2010, which has been applicable law since 2 July 2010. On October 12, 2011, the French

National Assembly voted a law forbidding the use of Bisphenol A in products aimed at less than 3 year old children for 2013, and 2014 for all food containers.

Germany

On 19 September 2008, the German Federal Institute for Risk Assessment (Bundesinstitut für Risikobewertung, BfR) stated that there was no reason to change the current risk assessment for bisphenol A on the basis of the Lang Study.

In October 2009, the German environmental organization Bund für Umwelt und Naturschutz Deutschland requested a ban on BPA for children's products, especially pacifiers, and products that make contact with food. In response, some manufacturers voluntarily removed the problematic pacifiers from the market.

Netherlands

On 6 November 2008, the Dutch Food and Consumer Product Safety Authority (VWA) stated in a newsletter that baby bottles made from polycarbonate plastic do not release measurable concentrations of bisphenol A and therefore are safe to use.

Switzerland

In February 2009, the Swiss Federal Office for Public Health, based on reports of other health agencies, stated that the intake of bisphenol A from food represents no risk to the consumer, including newborns and infants. However, in the same statement, it advised for proper use of polycarbonate baby bottles and listed alternatives.

Sweden

By May 2010, the Swedish Chemicals Agency asked for a BPA ban in baby bottles, but the Swedish Food Safety Authority prefers to await the expected European Food Safety Authority's updated review. The Minister of Environment said to wait for the EFSA review but not for too long. From March 2011 it is prohibited to manufacture babybottles containing bisphenol A and from July 2011 will no longer go to buy them in stores.

UK

In December 2009, responding to a letter from a group of seven scientists that urged the UK Government to "adopt a standpoint consistent with the approach taken by other Governments who have ended the use of BPA in food contact products marketed at children", the UK Food Standards Agency reaffirmed, in January 2009, its view that "exposure of UK consumers to BPA from all sources, including food contact materials, was well below levels considered harmful".

Turkey

As of 10 June 2011, Turkey banned the use of BPA in baby bottles and other PC items produced for babies.

Japan

Between 1998 and 2003, the canning industry voluntarily replaced their BPA-containing epoxy resin can liners with BPA-free polyethylene terephthalate (PET) in many of their products. For other products, they switched to a different epoxy lining that yielded much less migration of BPA into food than the previously used resin. In addition, polycarbonate tableware for school lunches was replaced by BPA-free plastics. As a result of these changes, Japanese risk assessors have found that virtually no BPA is detectable in canned foods or drinks, and blood levels of BPA in people have declined dramatically (50% in one study).

United States

September 2008

In September, the National Toxicology Program finalized their report on bisphenol A, finding "some concern", mid-point of a five-level scale, that infants were at risk from exposure to the chemical.

At that time, the FDA reassured consumers that current limits were safe, but convened an outside panel of experts to review the issue. The Lang study was also released that month, and David Melzer, a co-author of the study, presented the results of the study before the FDA panel.

The editorial accompanying the Lang study's publication in *JAMA* criticized the FDA's assessment of bisphenol A: "A fundamental problem is that the current ADI [acceptable daily intake] for BPA is based on experiments conducted in the early 1980s using outdated methods (only very high doses were tested) and insensitive assays.

More recent findings from independent scientists were rejected by the FDA, apparently because those investigators did not follow the outdated testing guidelines for environmental chemicals, whereas studies using the outdated, insensitive assays (predominantly involving studies funded by the chemical industry) are given more weight in arriving at the conclusion that BPA is not harmful at current exposure levels."

March 2009

Sunoco, a producer of gasoline and chemicals, is now refusing to sell the chemical to companies for use in food and water containers for children younger than 3, saying it can't be certain of the compound's safety. Sunoco plans to require its customers to guarantee that the chemical will not be used in children's food products.

The six largest US companies that produce baby bottles decided to stop using bisphenol A in their products. Suffolk County, New York banned baby beverage containers made with bisphenol A.

On 13 March, leaders from the House and Senate proposed legislation to ban bisphenol A. In the same month, Rochelle Tyl, author of two studies used by FDA to assert BPA safety in August 2008, said those studies didn't claim that BPA is safe because they weren't designed to cover all aspects of the chemical's effects.

May 2009

The first US jurisdictions to pass regulations limiting or banning BPA were Minnesota and Chicago. Minnesota's regulation takes effect in 2010, "manufacturers of... children's products containing BPA may not sell them in the state after 1 Jan. 2010. The ban extends to all retailers in the state a year later." The products impacted are known as sippy cups and baby bottles.

The City of Chicago adopted a similar ban shortly thereafter. Coverage of Chicago's ban in the news showed a relentless opposition by the industry. A Chicago Tribune article noted an up-hill battle while passing legislation, "[industry officials] used FDA's position on the issue when they tried to block the city's measure."

In May 2009, the Washington Post accused the manufacturers of food and beverage containers and some of their biggest customers of trying to devise a public relations and lobbying strategy to block government BPA bans. Lyndsey Layton, from the Washington Post, criticized the FDA noting that, "Despite more than 100 published studies by government scientists and university laboratories that have raised health concerns about the chemical, the Food and Drug Administration has deemed it safe largely because of two studies, both funded by a chemical industry trade group".

June 2009

In June 2009, the FDA announced the decision to reconsider the BPA safety levels.

Connecticut was the first US state to ban bisphenol A from infant formula and baby food containers, as well from any reusable food or beverage container.

July 2009

The California Environmental Protection Agency's Developmental and Reproductive Toxicant Identification Committee unanimously voted against placing Bisphenol A on the state's list of chemicals that are believed to cause reproductive harm. The panel, although concerned over the growing scientific research showing BPA's reproductive harm in animals, found that there was insufficient data of the effects in humans. Critics point out that the same panel failed to add second-hand smoke to the list until 2006, and only one chemical was added to the list in the last three years.

August 2009

On 3 August, Massachusetts' Department of Public Health advised mothers to take certain actions to prevent possible health impact in children. Mothers with children up to two years old were advised to limit exposure by avoiding products that might contain BPA, such as plastic drinking bottles and other plastic materials with recycling codes of 7 or 3.

The *Milwaukee Journal Sentinel*, as part of an ongoing investigative series into BPA and its effects, revealed plans by the Society of the Plastics Industry to execute a major public relations blitz to promote BPA, including plans to attack and discredit those who report or comment negatively on the monomer and its effects.

September 2009

On 29 September, the U.S. Environmental Protection Agency announced that it is evaluating BPA, and another five chemicals, for action plan development.

October 2009

On 28 October, the NIH announced $30,000,000 in stimulus grants to study the health effects of BPA. This money is expected to result in many peer-reviewed publications.

November 2009

The *Consumer Reports* magazine published an analysis of BPA content in some canned foods and beverages, where in specific cases the content of a single can of food could exceed the current FDA Cumulative Exposure Daily Intake.

January 2010

On 15 January, the FDA expressed "some concern", the middle level in the scale of concerns, about the potential effects of BPA on the brain, behaviour, and prostate gland in fetuses, infants, and young children, and announced it was taking reasonable steps to reduce human exposure to BPA in the food supply. However, the FDA is not recommending that families change the use of infant formula or foods, as it sees the benefit of a stable source of good nutrition as outweighing the potential risk from BPA exposure.

On the same date, the U.S. Department of Health & Human Services released information to help parents to reduce children's BPA exposure.

February 2010

According to *The Milwaukee Journal Sentinel*, which supports a BPA ban, after lobbyists for the chemical industry met with administration officials, the EPA delayed BPA regulation and did not include the chemical in an action plan released 30 December 2009.

Many US states are considering some sort of BPA ban.

March 2010

On 29 March, the EPA declared BPA a "chemical of concern".

April 2010

The 2008–2009 Annual Report of the President's Cancer Panel declared: "because of the long latency period of many cancers, the available evidence argues for a precautionary approach to these diverse chemicals, which include (...) bisphenol A"

Meanwhile, as of April 2011, General Mills has announced that it has found a BPA-free alternative can liner that apparently works even with tomatoes, a highly acidic product that has long baffled the industry in terms of finding a suitable substitute. General Mills says that with the next tomato harvest, it will begin using the BPA-free alternative in tomato products sold by its organic foods subsidiary Muir Glen. Thus far, there has been no word on whether General Mills will use BPA-free alternatives on any of its other canned products.

February 2011

In August 2010, the Maine Board of Environmental Protection voted unanimously to ban the sale of baby bottles and other reusable food and beverage containers made with bisphenol A as of January 2012. In February 2011, the newly elected governor of Maine, Paul

LePage, gained national attention when he spoke on a local TV news show saying he hoped to repeal the ban because, "There hasn't been any science that identifies that there is a problem" and added: "The only thing that I've heard is if you take a plastic bottle and put it in the microwave and you heat it up, it gives off a chemical similar to estrogen. So the worst case is some women may have little beards." In April, the Maine legislature passed a bill to ban the use of BPA in baby bottles, sippy cups, and other reusable food and beverage containers, effective January 1, 2012. Governor LePage refused to sign the bill.

Polycarbonate

Polycarbonates, known by the trademarked names Lexan, Makrolon, Makroclear and others, are a particular group of thermoplastic polymers. They are easily worked, moulded, and thermoformed. Because of these properties, polycarbonates find many applications. Polycarbonates do not have a unique plastic identification code and are identified as Other, 7.

Structure

Polycarbonates received their name because they are polymers containing carbonate groups (–O–(C=O)–O–). Most polycarbonates of commercial interest are derived from rigid monomers. A balance of useful features including temperature resistance, impact resistance and optical properties position polycarbonates between commodity plastics and engineering plastics.

Production

The main polycarbonate material is produced by the reaction of bisphenol A and phosgene $COCl_2$. The overall reaction can be written as follows:

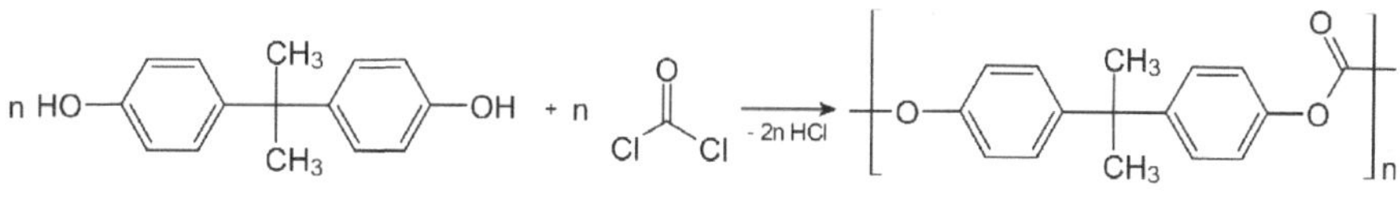

The first step of the synthesis involves treatment of bisphenol A with sodium hydroxide, which deprotonates the hydroxyl groups of the bisphenol A.

$(HOC_6H_4)_2CMe_2 + 2\ NaOH \rightarrow (NaOC_6H_4)_2CMe_2 + 2\ H_2O$

The diphenoxide ($(NaOC_6H_4)_2CMe_2$) reacts with phosgene to give a chloroformate, which subsequently is attacked by another phenoxide. The net reaction from the diphenoxide is:

$$(NaOC_6H_4)_2CMe_2 + COCl_2 \rightarrow 1/n\ [OC(OC_6H_4)_2CMe_2]_n + 2\ NaCl$$

In this way, approximately one billion kilograms of polycarbonate is produced annually. Many other diols have been tested in place of bisphenol A, e.g. 1,1-bis(4-hydroxyphenyl)cyclohexane and dihydroxybenzophenone including some, e.g. tetramethylcyclobutanediol, that are unlikely endocrine disruptors.

An alternative route to polycarbonates entails transesterification from BPA and diphenyl carbonate:

$$(HOC_6H_4)_2CMe_2 + (C_6H_5O)_2CO \rightarrow 1/n\ [OC(OC_6H_4)_2CMe_2]_n + 2\ C_6H_5OH$$

The diphenyl carbonate was derived in part from carbon monoxide, this route being greener than the phosgene method.

Properties and Processing

Polycarbonate is a very durable material. Although it has high impact-resistance, it has low scratch-resistance and so a hard coating is applied to polycarbonate eyewear lenses and polycarbonate exterior automotive components. The characteristics of polycarbonate are quite like those of polymethyl methacrylate (PMMA, acrylic), but polycarbonate is stronger, usable in a wider temperature range but more expensive. This polymer is highly transparent to visible light and has better light transmission characteristics than many kinds of glass.

Polycarbonate has a glass transition temperature of about 150 °C (302 °F), so it softens gradually above this point and flows above about 300 °C (572 °F). Tools must be held at high temperatures, generally above 80 °C (176 °F) to make strain- and stress-free products. Low molecular mass grades are easier to mould than higher grades, but their strength is lower as a result. The toughest grades have the highest molecular mass, but are much more difficult to process.

Unlike most thermoplastics, polycarbonate can undergo large plastic deformations without cracking or breaking. As a result, it can be processed and formed at room temperature using sheet metal techniques, such as forming bends on a brake.

Even for sharp angle bends with a tight radius, no heating is generally necessary. This makes it valuable in prototyping applications where transparent or electrically non-conductive parts are needed, which cannot be made from sheet metal. Note that PMMA/Plexiglas, which is similar in appearance to polycarbonate, is brittle and cannot be bent at room temperature.

Main transformation techniques for polycarbonate resins:

- extrusion into tubes, rods and other profiles
- extrusion with cylinders into sheets (0.5–15 mm (0.020–0.59 in)) and films (below 1 mm (0.039 in)), which can be used directly or manufactured into other shapes using thermoforming or secondary fabrication techniques, such as bending, drilling, routing, laser cutting etc.
- injection molding into ready articles

Applications

Electronic Components

Polycarbonate is mainly used for electronic applications that capitalize on its collective safety features. Being a good electrical insulator and having heat resistant and flame retardant properties, it is used in various products associated with electrical and telecommunications hardware. It also serves as dielectric in high stability capacitors.

Construction Materials

The second largest consumer of polycarbonates is the construction industry, e.g. for domelights, flat or curved glazing, and sound walls.

Data Storage

A major application of polycarbonate is the production of Compact Discs, DVDs, and Blu-ray Discs. These discs are produced by injection molding polycarbonate into a mold cavity that has on one side a metal stamper containing a negative image of the disc data, while the other mold side is a mirrored surface. Typical products of sheet/film production include applications in advertisement (signs, displays, poster protection).

Automotive, Aircraft, and Security Components

In the automotive industry, injection-moulded polycarbonate can produce very smooth surfaces that make it well-suited for direct (without the need for a basecoat) metalised parts such as decorative bezels and optical reflectors.

Its uniform mould shrinkage results in parts with greater accuracy than those made of polypropylene. However, due to its susceptibility to environmental stress cracking, its use is limited to low-stress applications. It can be laminated to make bullet-proof "glass", although "bullet-resistant" is more accurate for the thinner windows, such as are used in bullet-resistant windows in automobiles. The thicker barriers of transparent plastic used in teller's windows and barriers in banks are also polycarbonate.

So-called "theft-proof" large plastic packaging for smaller items, which cannot be opened by hand, is uniformly made from polycarbonate.

The cockpit canopy of the F-22 Raptor jet fighter is made from a piece of high optical quality polycarbonate, and is the largest piece of its type formed in the world.

Niche Applications

Polycarbonate, being a versatile material with attractive processing and physical properties, has attracted myriad smaller applications. The use of injection molded drinking bottles and glasses and food containers has stirred serious controversy.

Many kinds of lenses are manufactured from polycarbonate, including durable, lightweight luggage, automotive headlamp lenses, lighting lenses, sunglass/eyeglass, lenses, and safety glasses.

Other miscellaneous items: MP3/digital audio player cases, ocarinas, computer cases, riot shields, visors, instrument panels, and blender jars. Many toys and hobby items are made from polycarbonate parts, e.g. fins, gyro mounts, and flybar locks for use with radio-controlled helicopters.

For use in applications exposed to weathering or UV-radiation, a special surface treatment is needed. This either can be a coating (e.g. for improved abrasion resistance), or a coextrusion for enhanced weathering resistance.

Polycarbonate is also used as a printing substrate for nameplate and other forms of industrial grade under printed products. The polycarbonate provides a barrier to wear, the elements, and fading.

Medical Applications

Some polycarbonate grades are used in medical applications and comply with both ISO 10993-1 and USP Class VI standards (occasionally referred to as PC-ISO). Class VI is the most stringent of the six USP ratings.

These grades can be sterilized using steam at 120 °C, gamma radiation, or by the ethylene oxide (EtO) method. However, scientific research indicates possible problems with biocompatibility. Dow Chemical strictly limits all its plastics with regard to medical applications.

Potential Hazards in Food Contact Applications

The use of polycarbonate containers for the purpose of food storage is controversial. The basis of this controversy is their hydrolysis (degradation by water, often referred to as leaching) releases bisphenol A:

$$1/n\ [OC(OC_6H_4)_2CMe_2]_n + H_2O \rightarrow (HOC_6H_4)_2CMe_2 + CO_2$$

More than 100 studies have explored the bioactivity of bisphenol A derived from polycarbonates. Bisphenol A appeared to be released from polycarbonate animal cages into water at room temperature and it may have been responsible for enlargement of the reproductive organs of female mice.

However, the animal cages used in the research were fabricated from industrial grade polycarbonate, rather than FDA food grade polycarbonate.

An analysis of the literature on bisphenol A leachate low-dose effects by vom Saal and Hughes published in August 2005 seems to have found a suggestive correlation between the source of funding and the conclusion drawn. Industry funded studies tend to find no significant effects whereas government funded studies tend to find significant effects.

Sodium hypochlorite bleach and other alkali cleaners catalyze the release of the bisphenol A from polycarbonate containers. A chemical compatibility chart shows that polycarbonate is incompatible with ammonia and acetone because it dissolves in their presence. Alcohol is one recommended organic solvent for cleaning grease and oils from polycarbonate.

Phosgene

Phosgene is the chemical compound with the formula $COCl_2$. This colorless gas gained infamy as a chemical weapon during World War I. It is also a valued industrial reagent and building block in synthesis of pharmaceuticals and other organic compounds. In low concentrations, its odor resembles freshly cut hay or grass.

In addition to its industrial production, small amounts occur naturally from the breakdown and the combustion of organochlorine compounds, such as those used in refrigeration systems. The name, sounding similar to "phosphine", does not mean it has any phosphorus. The chemical was named by combining the Greek words 'phos' (meaning light) and genesis (birth).

Structure and Basic Properties

Phosgene is a planar molecule as predicted by VSEPR theory. The C=O distance is 1.18 Å, the C—Cl distance is 1.74 Å and the Cl—C—Cl angle is 111.8°. It is one of the simplest acid chlorides, being formally derived from carbonic acid.

Production

Industrially, phosgene is produced by passing purified carbon monoxide and chlorine gas through a bed of porous activated carbon, which serves as a catalyst:

$$CO + Cl_2 \rightarrow COCl_2 \ (\Delta H_{rxn} = -107.6 kJ/mol)$$

The reaction is exothermic, therefore the reactor must be cooled. Typically, the reaction is conducted between 50 and 150 °C. Above 200 °C, phosgene reverts to carbon monoxide and chlorine, K_{eq} (300K) = 0.05. Approximately 5000 tonnes were produced in 1989.

Because of safety issues, phosgene is almost always produced and consumed within the same plant and extraordinary measures are made to contain this toxic gas. It is listed on schedule 3 of the Chemical Weapons Convention: All production sites manufacturing more than 30 tonnes per year must be declared to the OPCW.

Although less dangerous than many other chemical weapons, such as sarin, phosgene is still regarded as a viable chemical warfare agent because it is so easy to manufacture when compared to the production requirements of more technically advanced chemical weapons such as the first-generation nerve agent tabun.

Adventitious Occurrence

Upon ultraviolet (UV) radiation in the presence of oxygen, chloroform slowly converts into phosgene via a radical reaction. To suppress this photodegradation, chloroform is often stored in brown-tinted glass containers. Chlorinated compounds used to remove oil from metals, such as automotive brake cleaners, are converted to phosgene by the UV rays of arc welding processes.

Phosgene may also be produced during testing for leaks of older-style refrigerant gasses. Chloromethanes (R12, R22 and others) were formerly leak-tested in situ by employing a small gas torch (propane, butane or propylene gas) with a sniffer tube and a copper reaction plate in the flame nozzle of the torch.

If any refrigerant gas was leaking from a pipe or joint, the gas would be sucked into the flame via the sniffer tube and would cause a colour change of the gas flame to a bright greenish blue. In the process, phosgene gas would be created due to the thermal reaction.

No valid statistics are available, but anecdotal reports suggest that numerous refrigeration technicians suffered the effects of phosgene poisoning due to their ignorance of the toxicity of phosgene, produced during such leak testing.

Electronic sensing of refrigerant gases phased out the use of flame testing for leaks in the 1980s. Similarly, phosgene poisoning is a consideration for people fighting fires that are occurring in the vicinity of freon refrigeration equipment, smoking in the vicinity of a freon leak, or fighting fires using halon or halotron.

Uses

The great majority of phosgene is used in the production of isocyanates, the most important being toluene diisocyanate (TDI) and methylene diphenyl diisocyanate (MDI). These isocyanates are precursors to polyurethanes. Significant amounts are also used in the production of polycarbonates via its reaction with bisphenol A. Polycarbonates are an important class of engineering thermoplastic found, for example, in lenses in eye glasses.

Organic Synthesis

Although phosgene still finds use in organic synthesis, a variety of substitutes have been developed, notably trichloromethyl chloroformate ("diphosgene"), which is a liquid at room temperature, and bis(trichloromethyl) carbonate ("triphosgene"), a crystalline substance. The following are the three most useful reactions involving phosgene.

Synthesis of Carbonates

Diols react with phosgene to give either linear or cyclic carbonates (R = H, alkyl, aryl):

$$HOCR_2\text{-}X\text{-}CR_2OH + COCl_2 \rightarrow 1/n\ [OCR_2\text{-}X\text{-}CR_2OC(O)\text{-}]_n + 2\ HCl$$

Synthesis of Isocyanates

The synthesis of isocyanates from amines illustrates the electrophilic character of this reagent and its use in introducing the equivalent of "CO^{2+}" (R = alkyl, aryl):

$$RNH_2 + COCl_2 \rightarrow RN{=}C{=}O + 2\ HCl$$

Such reactions are conducted in the presence of a base such as pyridine that absorbs the hydrogen chloride.

Synthesis of Acid Chlorides

It is also used to produce acid chlorides and carbon dioxide from carboxylic acids:

$$RCO_2H + COCl_2 \rightarrow RC(O)Cl + HCl + CO_2$$

Such acid chlorides react with amines and alcohols to give, respectively, amides and esters, which are commonly used intermediates. Thionyl chloride is more commonly and more safely employed for this application. A specific application for phosgene is the production of chloroformic esters:

$$ROH + COCl_2 \rightarrow ROC(O)Cl + HCl$$

Inorganic Chemistry

Although it is somewhat hydrophobic, phosgene reacts with water to release hydrogen chloride and carbon dioxide:

$$COCl_2 + H_2O \rightarrow CO_2 + 2\ HCl$$

Analogously, with ammonia, one obtains urea:

$$COCl_2 + 4\ NH_3 \rightarrow CO(NH_2)_2 + 2\ NH_4Cl$$

Halide exchange with nitrogen trifluoride and aluminium tribromide gives COF_2 and $COBr_2$, respectively.

History

Phosgene was synthesized by the British chemist John Davy (1790–1868) in 1812 by exposing a mixture of carbon monoxide and chlorine to sunlight. He named it "phosgene" in reference of the use of light to promote the reaction; from Greek, *phos* (light) and *gene* (born). It gradually became important in the chemical industry as the 19th century progressed, particularly in dye manufacturing.

Chemical Warfare

Following the extensive use of phosgene gas in combat during World War I, it was stockpiled by various countries as part of their secret chemical weapons programs.

Phosgene was then only frequently used by the Imperial Japanese Army against the Chinese during the Second Sino-Japanese War. Gas weapons, such as phosgene, were produced by Unit 731 and authorized by specific orders given by Hirohito (Emperor Showa) himself, transmitted by the chief of staff of the army.

For example, the Emperor authorized the use of toxic gas on 375 separate occasions during the battle of Wuhan from August to October 1938.

Safety

Phosgene is an insidious poison as the odor may not be noticed and symptoms may be slow to appear. Phosgene can be detected at 0.4 ppm, which is four times the Threshold Limit Value. Its high toxicity arises from the action of the phosgene on the proteins in the pulmonary alveoli, which are the site of gas exchange: their damage disrupts the blood-air barrier, causing suffocation.

It reacts with the amines of the proteins, causing crosslinking via formation of urea-like linkages, in accord with the reactions. Phosgene detection badges are worn by those at risk of exposure.

Sodium bicarbonate may be used to neutralise liquid spills of phosgene. Gaseous spills may be mitigated with ammonia.

Solvent

A solvent (from the Latin *solvô*, "I loosen, untie, I solve") is a liquid, solid, or gas that dissolves another solid, liquid, or gaseous solute, resulting in a solution that is soluble in a certain volume of solvent at a specified temperature. Common uses for organic solvents are in dry cleaning (e.g., tetrachloroethylene), as a paint thinner (e.g., toluene, turpentine), as nail polish removers and glue solvents (acetone, methyl acetate, ethyl acetate), in spot removers (e.g., hexane, petrol ether), in detergents (citrus terpenes), in perfumes (ethanol), and nail polish, (part of how it stays on nails) and in chemical synthesis. The use of inorganic solvents (other than water) is typically limited to research chemistry and some technological processes.

In 2005, the worldwide market for solvents had a total volume of around 17.9 million tons, which led to a turnover of about €8,000,000,000.

Solutions and Solvation

When one substance is dissolved into another, a solution is formed. This is opposed to the situation when the compounds are insoluble and one of them precipitate like sand in water. In solution, all of the ingredients are uniformly distributed at a molecular level and no residue remains. The mixing is referred to as miscibility, whereas the ability to dissolve one compound into another is known as solubility. However, in addition to mixing, both substances in the solution interact with each other.

When something is dissolved, molecules of the solvent arrange themselves around molecules of the solute. Heat is involved and entropy is increased making the solution more thermodynamically stable than the solute alone. This arrangement is mediated by the respective chemical properties of the solvent and solute, such as hydrogen bonding, dipole moment and polarizability.

Solvent Classifications

Solvents can be broadly classified into two categories: *polar* and *non-polar*. Generally, the dielectric constant of the solvent provides a rough measure of a solvent's polarity. The strong polarity of water is indicated, at 20 °C, by a dielectric constant of 80.10;. Solvents with a dielectric constant of less than 15 are generally considered to be nonpolar.

Technically, the dielectric constant measures the solvent's ability to reduce the field strength of the electric field surrounding a charged particle immersed in it. This reduction is then compared to the field strength of the charged particle in a vacuum. In laymen's terms, dielectric constant of a solvent can be thought of as its ability to reduce the solute's internal charge.

Other Polarity Scales

Dielectric constants are not the only measure of polarity. Because solvents are used by chemists to carry out chemical reactions or observe chemical and biological phenomena, more specific measures of polarity are required.

The Grunwald Winstein mY scale measures polarity in terms of solvent influence on buildup of positive charge of a solute during a chemical reaction.

Kosower's Z scale measures polarity in terms of the influence of the solvent on uv absorption maxima of a salt, usually pyridinium iodide or the pyridinium zwitterion.

Donor number and donor acceptor scale measures polarity in terms of how a solvent interacts with specific substances, like a strong Lewis acid or a strong Lewis base.

The polarity, dipole moment, polarizability and hydrogen bonding of a solvent determines what type of compounds it is able to dissolve and with what other solvents or liquid compounds it is miscible. As a rule of thumb, polar solvents dissolve polar compounds best and non-polar solvents dissolve non-polar compounds best: "like dissolves like". Strongly polar compounds like sugars (e.g., sucrose) or ionic compounds, like inorganic salts (e.g., table salt) dissolve only in very polar solvents like water, while strongly non-polar compounds like oils or waxes dissolve only in very non-polar organic solvents like hexane. Similarly, water and hexane (or vinegar and vegetable oil) are not miscible with each other and will quickly separate into two layers even after being shaken well.

Polar Protic and Polar Aprotic

Solvents with a relative static permittivity greater than 15 can be further divided into protic and aprotic. Protic solvents solvate anions (negatively charged solutes) strongly via hydrogen bonding. Water is a protic solvent. Aprotic solvents such as acetone or dichloromethane tend to have large dipole moments (separation of partial positive and partial negative charges within the same molecule) and solvate positively charged species via their negative dipole.

In chemical reactions the use of polar protic solvents favors the S_N1 reaction mechanism, while polar aprotic solvents favour the S_N2 reaction mechanism.

Physical Properties of Common Solvents

Properties Table of Common Solvents

The solvents are grouped into non-polar, polar aprotic, and polar protic solvents and ordered by increasing polarity. The polarity is given as the dielectric constant. The properties of solvents that exceed those of water are bolded.

Solvent	*Chemical formula*	*Boiling point*	*Dielectric constant*	*Density*	*Dipole moment*
	Non-polar solvents				
Pentane	CH_3-CH_2-CH_2-CH_2-CH_3	36 °C	1.84	0.626 g/ml	0.00 D
Cyclopentane	C_5H_{10}	40 °C	1.97	0.751 g/ml	0.00 D
Hexane	CH_3-CH_2-CH_2-CH_2-CH_2-CH_3	69 °C	1.88	0.655 g/ml	0.00 D
Cyclohexane	C_6H_{12}	81 °C	2.02	0.779 g/ml	0.00 D
Benzene	C_6H_6	80 °C	2.3	0.879 g/ml	0.00 D
Toluene	C_6H_5-CH_3	111 °C	2.38	0.867 g/ml	0.36 D
1,4-Dioxane	/-CH_2-CH_2-O-CH_2-CH_2-O-\	101 °C	2.3	1.033 g/ml	0.45 D
Chloroform	$CHCl_3$	61 °C	4.81	1.498 g/ml	1.04 D
Diethyl ether	CH_3CH_2-O-CH_2-CH_3	35 °C	4.3	0.713 g/ml	1.15 D

Contd...

Solvent	*Chemical formula*	*Boiling point*	*Dielectric constant*	*Density*	*Dipole moment*
	Polar aprotic solvents				
Dichloromethane (DCM)	CH_2Cl_2	40 °C	9.1	1.3266 g/ml	1.60 D
Tetrahydrofuran (THF)	/-CH_2-CH_2-O-CH_2-CH_2-\	66 °C	7.5	0.886 g/ml	1.75 D
Ethyl acetate	CH_3-C(=O)-O-CH_2-CH_3	77 °C	6.02	0.894 g/ml	1.78 D
Acetone	CH_3-C(=O)-CH_3	56 °C	21	0.786 g/ml	2.88 D
Dimethylformamide (DMF)	H-C(=O)N$(CH_3)_2$	153 °C	38	0.944 g/ml	3.82 D
Acetonitrile (MeCN)	CH_3-Ca"N	82 °C	37.5	0.786 g/ml	3.92 D
Dimethyl sulfoxide (DMSO)	CH_3-S(=O)-CH_3	189 °C	46.7	1.092 g/ml	3.96 D
	Polar protic solvents				
Formic acid	H-C(=O)OH	101 °C	58	1.21 g/ml	1.41 D
n-Butanol	CH_3-CH_2-CH_2-CH_2-OH	118 °C	18	0.810 g/ml	1.63 D
Isopropanol (IPA)	CH_3-CH(-OH)-CH_3	82 °C	18	0.785 g/ml	1.66 D
n-Propanol	CH_3-CH_2-CH_2-OH	97 °C	20	0.803 g/ml	1.68 D
Ethanol	CH_3-CH_2-OH	79 °C	24.55	0.789 g/ml	1.69 D
Methanol	CH_3-OH	65 °C	33	0.791 g/ml	1.70 D
Acetic acid	CH_3-C(=O)OH	118 °C	6.2	1.049 g/ml	1.74 D
Water	H-O-H	100 °C	80	1.000 g/ml	1.85 D

There's another powerful way to look at these same solvents. By knowing their Hansen solubility parameter values (HSPiP), which are based on δD=dispersion bonds, δP=polar bonds and δH=hydrogen bonds, you know important things about their inter-molecular interactions with other solvents and also with polymers, pigments, nanoparticles etc. so you can do two things.

First, you can create rational formulations knowing, for example, that there is a good HSP match between a solvent and a polymer.

Second, you can make rational substitutions for "good" solvents (they dissolve things well) that are "bad" (for the environment, for health, for cost etc.).

The following table shows that the intuitions from "non-polar", "polar aprotic" and "polar protic" are put numerically - the "polar" molecules have higher levels of δP and the protic solvents have higher levels of δH. Because numerical values are used, comparisons can be made rationally by comparing numbers. So acetonitrile is much more polar than acetone but slightly less hydrogen bonding.

Solvent	***Chemical formula***	***dD Dispersion***	***dP Polar***	***dH Hydrogen bonding***
	Non-polar solvents			
Hexane	CH_3-CH_2-CH_2-CH_2-CH_2-CH_3	14.9	0.0	0.0
Benzene	C_6H_6	18.4	0.0	2.0
Toluene	C_6H_5-CH_3	18.0	1.4	2.0
Diethyl ether	CH_3CH_2-O-CH_2-CH_3	14.5	2.9	4.6
Chloroform	$CHCl_3$	17.8	3.1	5.7
1,4-Dioxane	/-CH_2-CH_2-O-CH_2-CH_2-O-\	17.5	1.8	9.0
	Polar aprotic solvents			
Ethyl acetate	CH_3-C(=O)-O-CH_2-CH_3	15.8	5.3	7.2
Tetrahydrofuran (THF)	/-CH_2-CH_2-O-CH_2-CH_2-\	16.8	5.7	8.0
Dichloromethane	CH_2Cl_2	17.0	7.3	7.1
Acetone	CH_3-C(=O)-CH_3	15.5	10.4	7.0
Acetonitrile (MeCN)	CH_3-Ca"N	15.3	18.0	6.1
Dimethylformamide (DMF)	H-C(=O)N$(CH_3)_2$	17.4	13.7	11.3
Dimethyl sulfoxide (DMSO)	CH_3-S(=O)-CH_3	18.4	16.4	10.2
Polar protic solvents				
Acetic acid	CH_3-C(=O)OH	14.5	8.0	13.5
n-Butanol	CH_3-CH_2-CH_2-CH_2-OH	16.0	5.7	15.8
Isopropanol	CH_3-CH(-OH)-CH_3	15.8	6.1	16.4
n-Propanol	CH_3-CH_2-CH_2-OH	16.0	6.8	17.4
Ethanol	CH_3-CH_2-OH	15.8	8.8	19.4
Methanol	CH_3-OH	14.7	12.3	22.3
Formic acid	H-C(=O)OH	14.6	10.0	14.0
Water	H-O-H	15.5	16.0	42.3

Consider a simple example of rational substitution. Suppose for environmental reasons we needed to replace the chlorinated solvent, chloroform, with a solvent (blend) of equal solvency using a mixture of two non-chlorinated solvents from this table. Via trial-and-error, a spreadsheet or some software such as HSPiP we find that a 50:50 mix of toluene and 1,4 dioxane is a close match. The dD of the mixture is the average of 18.0 and 17.5 = 17.8.

The dP of the mixture is the average of 1.4 and 1.8 = 1.6 and the dH of the mixture is the average of 2.0 and 9.0 = 5.5. So the mixture is 17.8, 1.6, 5.5 compared to Chloroform at 17.8, 3.1, 5.7. Because Toluene itself has many health issues, other mixtures of solvents can be found using a full Hansen solubility parameter dataset.

Boiling Point

Solvent	*Boiling point (°C)*
ethylene dichloride	83.48
pyridine	115.25
methyl isobutyl ketone	116.5
methylene chloride	39.75
isooctane	99.24
carbon disulfide	46.3
carbon tetrachloride	76.75
o-xylene	144.42

An important property of solvents is boiling point. This also determines the speed of evaporation. Small amounts of low-boiling solvents like diethyl ether, dichloromethane, or acetone will evaporate in seconds at room temperature, while high-boiling solvents like water or dimethyl sulfoxide need higher temperatures, an air flow, or the application of vacuum for fast evaporation.

- Low boilers: boiling temperature below 100 °C (boiling point of water)
- Medium boilers: between 100 °C and 150 °C
- High boilers: above 150 °C

Density

Most organic solvents have a lower density than water, which means they are lighter and will form a separate layer on top of water. An important exception: most of the halogenated solvents like dichloromethane or chloroform will sink to the bottom of a container, leaving water as the top layer. This is important to remember when partitioning compounds between solvents and water in a separatory funnel during chemical syntheses.

Often, specific gravity is cited in place of density. Specific gravity is defined as the density of the solvent divided by the density of water at the same temperature.

As such, specific gravity is a unitless value. It readily communicates whether a water-insoluble solvent will float (SG < 1.0) or sink (SG > 1.0) when mixed with water.

Solvent	***Specific gravity***
Pentane	0.626
Petroleum ether	0.656
Hexane	0.659
Heptane	0.684
Diethyl amine	0.707
Diethyl ether	0.713
Triethyl amine	0.728
Tert-butyl methyl ether	0.741
Cyclohexane	0.779
Tert-butyl alcohol	0.781
Isopropanol	0.785
Acetonitrile	0.786
Ethanol	0.789
Acetone	0.790
Methanol	0.791
Methyl isobutyl ketone	0.798
Isobutyl alcohol	0.802
1-Propanol	0.803
Methyl ethyl ketone	0.805
2-Butanol	0.808
Isoamyl alcohol	0.809
1-Butanol	0.810
Diethyl ketone	0.814
1-Octanol	0.826
p-Xylene	0.861
m-Xylene	0.864
Toluene	0.867
Dimethoxyethane	0.868
Benzene	0.879

Contd...

Butyl acetate	0.882
1-Chlorobutane	0.886
Tetrahydrofuran	0.889
Ethyl acetate	0.895
o-Xylene	0.897
Hexamethylphosphorus triamide	0.898
2-Ethoxyethyl ether	0.909
N,N-Dimethylacetamide	0.937
Diethylene glycol dimethyl ether	0.943-
N,N-Dimethylformamide	0.944
2-Methoxyethanol	0.965
Pyridine	0.982
Propanoic acid	0.993
Water	1.000
2-Methoxyethyl acetate	1.009
Benzonitrile	1.01
1-Methyl-2-pyrrolidinone	1.028
Hexamethylphosphoramide	1.03
1,4-Dioxane	1.033
Acetic acid	1.049
Acetic anhydride	1.08
Dimethyl sulfoxide	1.092
Chlorobenzene	1.1066
Deuterium oxide	1.107
Ethylene glycol	1.115
Diethylene glycol	1.118
Propylene carbonate	1.21
Formic acid	1.22
1,2-Dichloroethane	1.245
Glycerin	1.261
Carbon disulfide	1.263
1,2-Dichlorobenzene	1.306
Methylene chloride	1.326

Contd...

Nitromethane	1.382
2,2,2-Trifluoroethanol	1.393
Chloroform	1.498
1,1,2-Trichlorotrifluoroethane	1.575
Carbon tetrachloride	1.594
Tetrachloroethylene	1.623

Health and Safety

Fire

Most organic solvents are flammable or highly flammable, depending on their volatility. Exceptions are some chlorinated solvents like dichloromethane and chloroform. Mixtures of solvent vapours and air can explode. Solvent vapours are heavier than air; they will sink to the bottom and can travel large distances nearly undiluted. Solvent vapours can also be found in supposedly empty drums and cans, posing a flash fire hazard; hence empty containers of volatile solvents should be stored open and upside down.

Both diethyl ether and carbon disulfide have exceptionally low autoignition temperatures which increase greatly the fire risk associated with these solvents. The autoignition temperature of carbon disulfide is below 100°C (212°F), so objects such as steam pipes, light bulbs, hotplates and recently extinguished bunsen burners are able to ignite its vapours.

Explosive Peroxide Formation

Ethers like diethyl ether and tetrahydrofuran (THF) can form highly explosive organic peroxides upon exposure to oxygen and light, THF is normally more able to form such peroxides than diethyl ether. One of the most susceptible solvents is diisopropyl ether.

The heteroatom (oxygen) stabilizes the formation of a free radical which is formed by the abstraction of a hydrogen atom by another free radical. The carbon centred free radical thus formed is able to react with an oxygen molecule to form a peroxide compound.

A range of tests can be used to detect the presence of a peroxide in an ether; one is to use a combination of iron sulfate and potassium thiocyanate. The peroxide is able to oxidize the Fe^{2+} ion to an Fe^{3+} ion which then form a deep red coordination complex with the thiocyanate.

In extreme cases the peroxides can form crystalline solids within the vessel of the ether.

Unless the desiccant used can destroy the peroxides, they will concentrate during distillation due to their higher boiling point. When sufficient peroxides have formed, they can form a crystalline and shock sensitive solid precipitate.

When this solid is formed at the mouth of the bottle, turning the cap may provide sufficient energy for the peroxide to detonate. Peroxide formation is not a significant problem when solvents are used up quickly; they are more of a problem for laboratories which take years to finish a single bottle. Ethers have to be stored in the dark in closed canisters in the presence of stabilizers like butylated hydroxytoluene (BHT) or over sodium hydroxide.

Peroxides may be removed by washing with acidic iron(II) sulfate, filtering through alumina, or distilling from sodium/benzophenone. Alumina does not destroy the peroxides; it merely traps them. The advantage of using sodium/benzophenone is that moisture and oxygen is removed as well.

Health Effects

General health hazards associated with solvent exposure include toxicity to the nervous system, reproductive damage, liver and kidney damage, respiratory impairment, cancer, and dermatitis.

Many solvents can lead to a sudden loss of consciousness if inhaled in large amounts. Solvents like diethyl ether and chloroform have been used in medicine as anesthetics, sedatives, and hypnotics for a long time. Ethanol (grain alcohol) is a widely used and abused psychoactive drug. Diethyl ether, chloroform, and many other solvents (e.g., from gasoline or glues) are used recreationally in glue sniffing, often with harmful long term health effects like neurotoxicity or cancer. Methanol can cause permanent blindness and death. It is also dangerous because it burns with an invisible flame.

It is interesting to note that ethanol has a synergistic effect when taken in combination with many solvents. For instance a combination of toluene/benzene and ethanol causes greater nausea/vomiting than either substance alone.

Some solvents including chloroform and benzene (an ingredient of gasoline) are carcinogenic. Many others can damage internal organs like the liver, the kidneys, or the brain.

Chronic exposure to organic solvents in the work environment can produce a range of adverse neuropsychiatric effects and, occupational exposure to organic solvents has been associated with higher numbers of painters regularly exposed to high levels of organic solvents suffering from alcoholism.

General Precautions

- Avoid being exposed to solvent vapours by working in a fume hood, or with local exhaust ventilation (LEV), or in a well ventilated area
- Keep the storage containers tightly closed
- Never use open flames near flammable solvents; use electrical heating instead
- Never flush flammable solvents down the drain; read safety data sheets for proper disposal information
- Avoid the inhalation of solvent vapours
- Avoid contact of the solvent with the skin — many solvents are easily absorbed through the skin. They also tend to dry the skin and may cause sores and wounds.

Environmental Contamination

A major pathway to induce health effects arises from spills or leaks of solvents that reach the underlying soil. Since solvents readily migrate substantial distances, the creation of widespread soil contamination is not uncommon; there may be about 5000 sites worldwide that have major subsurface solvent contamination; this is particularly a health risk if aquifers are affected.

Cyclohexane

Cyclohexane is a cycloalkane with the molecular formula C_6H_{12}. Cyclohexane is used as a nonpolar solvent for the chemical industry, and also as a raw material for the industrial production of adipic acid and caprolactam, both of which are intermediates used in the production of nylon.

On an industrial scale, cyclohexane is produced by reacting benzene with hydrogen. Because of its unique chemical and conformational properties, cyclohexane is also used in labs in analysis and as a standard.

Cyclohexane has a distinctive detergent-like odor, reminiscent of cleaning product (in which it is sometimes used).

Structure and Physical Properties

Conformation

The 6-vertexed ring does not conform to the shape of a perfect hexagon. The conformation of a flat 2D planar hexagon has considerable angle strain because its bonds are not 109.5 degrees; the torsional strain would also be considerable because all of the bonds would be eclipsed bonds. Therefore, to reduce torsional strain, cyclohexane adopts a three-dimensional structure known as the chair conformation. This was first proposed as early as 1890 by Hermann Sachse, but only gained widespread acceptance much later.

The new conformation puts the carbons at an angle of 109.5°. Half of the hydrogens are in the plane of the ring (*equatorial*) while the other half are perpendicular to the plane (*axial*). This conformation allows for the most stable structure of cyclohexane. Another conformation of cyclohexane exists, known as boat conformation, but it interconverts to the slightly more stable chair formation.

If cyclohexane is mono-substituted with a large substituent, then the substituent will most likely be found attached in an equatorial position, as this is the slightly more stable conformation.

Cyclohexane has the lowest angle and torsional strain of all the cycloalkanes, as a result cyclohexane has been deemed a 0 in total ring strain, a combination of angle and torsional strain. This also makes cyclohexane the most stable of the cycloalkanes and therefore will produce the least amount of heat (per CH_2 unit) when burned compared to the other cycloalkanes.

Solid Phases

Cyclohexane has two crystalline phases. The high-temperature phase I, stable between 186 K and the melting point 280 K, is a plastic crystal, which means the molecules retain some rotational degree of freedom. The low-temperature phase II is ordered. A metastable phase III has been obtained by rapid quenching.

Chemistry—Reactions

Pure cyclohexane in itself is rather unreactive, being a non-polar, hydrophobic hydrocarbon. It can react with very strong acids such as the superacid system HF + SbF_5 which will cause forced protonation and "hydrocarbon cracking". Substituted cyclohexanes, however, may be reactive under a variety of conditions, many of which are important to organic chemistry. Cyclohexane is highly flammable.

Derivatives

The specific arrangement of functional groups in cyclohexane derivatives, and indeed in most cycloalkane molecules, is extremely important in chemical reactions, especially reactions involving nucleophiles. Substituents on the ring must be in the axial formation to react with other molecules.

For example, the reaction of bromocyclohexane and a common nucleophile, a hydroxide anion (OH^-), would result in cyclohexene:

$$C_6H_{11}Br + OH^- \rightarrow C_6H_{10} + H_2O + Br^-$$

This reaction, commonly known as an elimination reaction or dehalogenation (specifically E2), requires that the bromine substituent be in the axial formation, opposing another axial H atom to react.

Assuming that the bromocyclohexane was in the appropriate formation to react, the E2 reaction would commence as such:

- The electron pair bond between the C-Br moves to the Br, forming Br^- and setting it free from cyclohexane
- The nucleophile (-OH) gives an electron pair to the adjacent axial H, setting H free and bonding to it to create H_2O
- The electron pair bond between the adjacent axial H moves to the bond between the two C-C making it C=C

Note: All three steps happen simultaneously, characteristic of all E2 reactions.

The reaction above will generate mostly E2 reactions and as a result the product will be mostly (~70%) cyclohexene. However, the percentage varies with conditions, and generally, two different reactions (E2 and S_N2) compete. In the above reaction, an S_N2 reaction would substitute the bromine for a hydroxyl (OH^-) group instead, but once again, the Br must be in axial to react. Once the S_N2 substitution is complete, the newly substituted OH group would flip back to the more stable equatorial position quickly (~1 millisecond).

Uses

Commercially most of cyclohexane produced is converted into cyclohexanone–cyclohexanol mixture (or "*KA oil*") by catalytic oxidation. KA oil is then used as a raw material for adipic acid and caprolactam. Practically, if the cyclohexanol content of KA oil is higher than cyclohexanone, it is more likely (economical) to be converted into adipic acid, and the reverse case, caprolactam production is more likely.

Such ratio in KA oil can be controlled by selecting suitable oxidation catalyst. Some of cyclohexane is used as an organic solvent.

Although much is already known about this cyclic hydrocarbon, research is still being done on cyclohexane and benzene mixtures and solid phase cyclohexane to determine hydrogen yields of the mix when irradiated at “195 °C.

Cyclohexane is also used for calibration of differential scanning calorimetry (DSC) instruments, because of a convenient crystal-crystal transition at “87.1 C.

Heat treating equipment manufacturer Surface Combustion uses cyclohexane as a carbon carrying gas in their high purity vacuum carburizing furnaces.

History

Unlike compounds like benzene, cyclohexane cannot easily be obtained from natural resources such as coal. Toward the end of the nineteenth century early chemical investigators had to depend on organic synthesis.

It took them 30 years to flesh out the details. In 1867 Marcellin Berthelot reduced benzene with hydroiodic acid at elevated temperatures.

He incorrectly identified the reaction product as n-hexane not only because of the convenient match in boiling point (69°C) but also because he did not believe benzene was a cyclic molecule (like his contemporary August Kekulé) but rather some sort of association of acetylene.

In 1870 one of his sceptics Adolf von Baeyer repeated the reaction and pronounced the same reaction product hexahydrobenzene and in 1890 Vladimir Markovnikov believed he was able to distill the same compound from Caucasus petroleum calling his concoction hexanaphtene.

In 1894 Baeyer synthesized cyclohexane starting with a Dieckmann condensation of pimelic acid followed by multiple reductions:

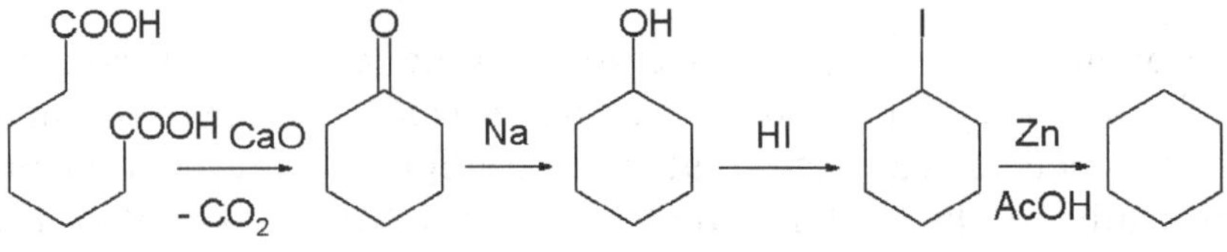

and in the same year E. Haworth and W.H. Perkin Jr. (1860–1929) did the same in a Wurtz reaction of 1,6-dibromohexane.

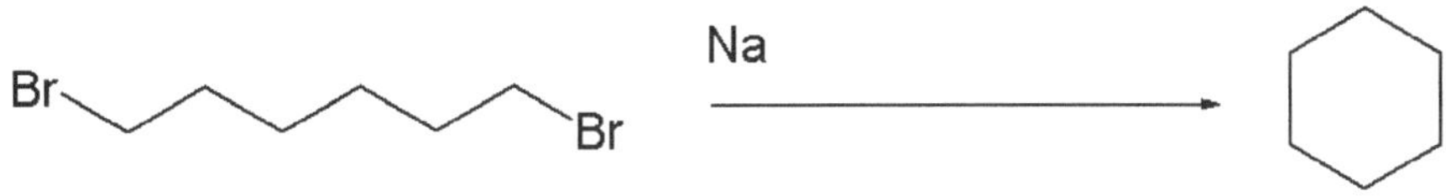

Surprisingly their cyclohexanes boiled higher by 10°C than either hexahydrobenzene or hexanaphtene but this riddle was solved in 1895 by Markovnikov, N.M. Kishner and Nikolay Zelinsky when they re-diagnosed hexahydrobenzene and hexanaphtene as methylcyclopentane, the result of an unexpected rearrangement reaction.

Chapter 2

Petroleum Technology

Hydraulic Fracturing

Hydraulic fracturing, often called fracking, fracing or hydrofracking, is the process of initiating and subsequently propagating a fracture in a rock layer, by means of a pressurized fluid, in order to release petroleum, natural gas, coal seam gas, or other substances for extraction. The fracturing, known colloquially as a frack job (or frac job), is done from a wellbore drilled into reservoir rock formations. The energy from the injection of a highly pressurized fluid, such as water, creates new channels in the rock which can increase the extraction rates and ultimate recovery of fossil fuels.

Hydraulic fractures may be natural or created by human activity, and are extended by internal fluid pressure which opens the fracture and causes it to extend through the rock. Natural hydraulic fractures include igneous dikes, sills and fracturing by ice as in frost weathering. Man-made fluid-driven fractures are formed at depth in a borehole and extend into targeted formations. The fracture width is typically maintained after the injection by introducing a *proppant* into the injected fluid. Proppant is a material, such as grains of sand, ceramic, or other particulates, that prevent the fractures from closing when the injection is stopped. The practice of hydraulic fracturing has come under scrutiny internationally due to concerns about environmental and health safety, and has been suspended or banned in some countries.

Purpose

The technique of hydraulic fracturing is used to increase or restore the rate at which fluids, such as oil or water, or natural gas can be produced from subterranean natural reservoirs, including unconventional reservoirs such as shale rock or coal beds. Hydraulic fracturing enables the production of natural gas and oil from rock formations deep below the earth's surface (generally 5,000-20,000 feet or 1,500-6,100 m). At such depth, there may not be sufficient porosity and permeability to allow natural gas and oil to flow from the rock into the wellbore at economic rates.

Thus, creating conductive fractures in the rock is essential to extract gas from shale reservoirs because of the extremely low natural permeability of shale, which is measured in the microdarcy to nanodarcy range. Fractures provide a conductive path connecting a larger area of the reservoir to the well, thereby increasing the area from which natural gas and liquids can be recovered from the targeted formation.

While the main industrial use of hydraulic fracturing is in stimulating production from oil and gas wells, hydraulic fracturing is also applied to:

- Stimulating groundwater wells
- Preconditioning rock for caving or inducing rock to cave in mining
- As a means of enhancing waste remediation processes, usually hydrocarbon waste or spills
- Dispose of waste by injection into deep rock formations
- As a method to measure the stress in the earth
- For heat extraction to produce electricity in an Enhanced Geothermal System.

Method

A hydraulic fracture is formed by pumping the fracturing fluid into the wellbore at a rate sufficient to increase pressure downhole to exceed that of the fracture gradient of the rock. The rock cracks and the fracture fluid continues farther into the rock, extending the crack still farther, and so on. To keep this fracture open after the injection stops, a solid proppant, commonly a sieved round sand, is added to the fluid. The propped fracture is permeable enough to allow the flow of formation fluids to the well. Formation fluids include gas, oil, salt water, fresh water and fluids introduced to the formation during completion of the well during fracturing.

The location of fracturing along the length of the borehole can be controlled by inserting composite plugs, also known as bridge plugs, above and below the region to be fractured. This allows a borehole to be progressively fractured along the length of the bore without fluid leaking through previously fractured regions. Fluid and proppant are introduced to the working region through piping in the upper plug. This method is commonly referred to as "plug and perf."

Typically, hydraulic fracturing is performed in cased wellbores and the zones to be fractured are accessed by perforating the casing at those locations.

Well Types

While hydraulic fracturing is many times performed in vertical wells, today it is more often performed in horizontal wells. Horizontal drilling involves wellbores where the terminal drillhole is completed as a 'lateral' that extends parallel with the rock layer containing the substance to be extracted. For example, laterals extend 1,500 to 5,000 feet in the Barnett Shale basin. In contrast, a vertical well only accesses the thickness of the rock layer, typically 50–300 feet. Horizontal drilling also reduces surface disruptions as fewer wells are required. Drilling a wellbore produces rock chips and fine rock particles that may enter cracks and pore space at the wellbore wall, reducing the permeability at and near the wellbore. This reduces flow into the borehole from the surrounding rock formation, and partially seals off the borehole from the surrounding rock. Hydraulic fracturing can be used to restore permeability.

Hydraulic fracturing is commonly applied to wells drilled in low permeability reservoir rock. An estimated 90 percent of the natural gas wells in the United States use hydraulic fracturing to produce gas at economic rates.

Fracturing

The fluid injected into the rock is typically a slurry of water, proppants, and chemical additives. Additionally, gels, foams, and compressed gases, including nitrogen, carbon dioxide and air can be injected. Various types of proppant include silica sand, resin-coated sand, and man-made ceramics. These vary depending on the type of permeability or grain strength needed. Sand containing naturally radioactive minerals is sometimes used so that the fracture trace along the wellbore can be measured. Chemical additives are applied to tailor the injected material to the specific geological situation, protect the well, and improve its operation, though the injected fluid is approximately more than 99 percent water and proppant, varying slightly based on the type of well. The composition of injected fluid is changed during the operation of a well over time, that is initially acid is used to increase permeability, then proppants are used with a gradual increase in their size and/or density, and at the end the well is flushed with water under pressure. Injected fluid is to some degree recovered and stored in pits or containers. Although the concentrations of the chemical additives are very low, the recovered fluid may be harmful due in part to hydrocarbons picked up from the formation. The recovered fluid is often processed so that part of it can be reused in another fracking operation, part released into the environment after treatment, and some left in the geologic formation.

Hydraulic fracturing equipment used in oil and natural gas fields usually consists of a slurry blender, one or more high pressure, high volume fracturing pumps (typically powerful triplex, or quintiplex pumps) and a monitoring unit. Associated equipment includes fracturing tanks, high pressure treating iron, a chemical additive unit (used to accurately monitor chemical addition) low pressure pipes and gauges for flow rate, fluid density, and treating pressure. Fracturing equipment operates over a range of pressures and injection rates, and can reach up to 100 MPa (15,000 psi) and 265 L/s (100 barrels per minute).

Fracture Monitoring

Microseismic monitoring is commonly used to estimate the size and orientation of hydraulically induced fractures. Microseismic activity is measured by placing an array of geophones in a nearby wellbore. By mapping the location of any small seismic events associated with the growing hydraulic fracture, the approximate geometry of the fracture is inferred. Tiltmeter arrays, deployed on the surface or down a well, provide another technology for monitoring the strains produced by hydraulic fracturing.

Emission of gases displaced by hydraulic fracturing into the atmosphere may be detected via atmospheric gas monitoring, and can be quantified directly via the eddy covariance flux measurements.

Completion

Advances in completion technology have led to the emergence of open hole multi-stage fracturing systems. These systems effectively place fractures in specific places in the wellbore, thus increasing the cumulative production in a shorter time frame. Certain reservoirs such as the Bakken, Barnett Shale, Montney and Haynesville Shale cannot be produced using conventional methods. These formations have begun using high tech completion systems capable of mechanically fracturing at certain intervals. An alternative to the plug and perf method, multi-stage fracturing systems are capable of stimulating several stages in a single day. Compared to the weeks required by the plug and perf method, cost-effective multi-stage completion systems are quickly becoming sought after technology by oil and natural gas companies.

Terminology

Fracture Gradient : The pressure to fracture the formation at a particular depth divided by the depth. A fracture gradient of 18 kPa/m (0.8 psi/foot) implies that at a depth of 3 km (10,000 feet) a pressure of 54 MPa (8,000 psi) will extend a hydraulic fracture.

ISIP - Initial Shut In Pressure : The pressure measured immediately after injection stops. The ISIP provides a measure of the pressure in the fracture at the wellbore by removing contributions from fluid friction.

Leakoff : Loss of fracturing fluid from the fracture channel into the surrounding permeable rock.

Fracturing fluid : The fluid used during a hydraulic fracture treatment of oil, gas or water wells. The fracturing fluid has two major functions:

1. Open and extend the fracture
2. Transport the proppant along the fracture length.

Proppant : Suspended particles in the fracturing fluid that are used to hold fractures open after a hydraulic fracturing treatment, thus producing a conductive pathway that fluids can easily flow along. Naturally occurring sand grains or artificial ceramic material are common proppants used.

Concise slang : "Fracing" (sometimes spelled "fracking" primarily in media) is a shortened version of fracturing.

Regulation

Hydraulic fracturing for the purpose of oil, natural gas, and geothermal production was exempted under the Safe Drinking Water Act. The decision for this may be due in part to the fact that underground sources of oil and gas already contain high levels of hydrocarbons, including carcinogens like benzene. Hydraulic fracturing for these purposes therefore falls under state jurisdiction. State regulations are inconsistent. All other hydraulic fracturing falls under the EPA regulations as do the handling and disposal of recovered fluids.

About 750 chemicals have been listed as additives for hydraulic fracturing in a report to the US Congress in 2011. Third-party laboratories are performing analysis on soil, air, and water near the fracturing sites to measure the level of contamination by each of the chemicals. Each state has a contact person in charge of such regulation. A map of these contact people can be found at FracFocus.org as well.

Environmental Concerns

A 2004 study by the Environmental Protection Agency (EPA) concluded that the injection of hydraulic fracturing fluids into CBM wells posed minimal threat to underground drinking water sources. This study has been criticised for only focusing on the injection of fracking fluids, while ignoring other aspects of the process such as disposal of fluids, and environmental concerns such as water quality, fish kills and acid burns; the study was also concluded before public complaints of contamination started emerging. Largely on the basis of this study, hydraulic fracturing was in 2005 exempted by US Congress from any regulation under the Safe Drinking Water Act.

With the explosive growth of natural gas wells in the US, researcher Valerie Brown predicted in 2007 that "public exposure to the many chemicals involved in energy development is expected to increase over the next few years, with uncertain consequences." As development of natural gas wells in the U.S. since the year 2000 has increased, so too have claims by private well owners of water contamination. This has prompted EPA and others to re-visit the topic.

Environmental concerns with hydraulic fracturing include the potential contamination of ground water, risks to air quality, the potential migration of gases and hydraulic fracturing chemicals to the surface, the potential mishandling of waste, and the health effects of these. While the EPA recognizes the potential for contamination of water by hydraulic fracturing, EPA Administrator Lisa Jackson testified in a Senate Hearing Committee "I'm not aware of any proven case where the fracking process itself has affected water...". There are, however, documented incidents of contamination. In 2006 drilling fluids and methane were detected leaking from the ground near a gas well in Clark, Wyoming; 8 million cubic feet of methane were eventually released, and shallow groundwater was found to be contaminated. In the town of Dimock, Pennsylvania, 13 water wells were contaminated with methane (one of them blew up), and the gas company, Cabot Oil & Gas, had to financially compensate residents and construct a pipeline to bring in clean water; the company continued to deny, however, that any "of the issues in Dimock have anything to do with hydraulic fracturing".

A Duke University study published in *Proceedings of the National Academy of Sciences* in 2011 examined methane in groundwater in Pennsylvania and New York states overlying the Marcellus Shale and the Utica Shale. It determined that groundwater tended to contain much higher concentrations of methane near fracking wells, with potential explosion hazard; the methane's isotopic signatures and other geochemical indicators were consistent with it originating in the fracked deep shale formations, rather than any other source. Complaints from a few residents on water quality in a developed natural gas field prompted an EPA groundwater investigation in Wyoming. The EPA reported detections of methane and other chemicals such as phthalates in private water wells.

In Pavillion, Wyoming, the EPA discovered traces of methane and foaming agents in several water wells near a gas rig, though it suggested these chemicals might have come from cleaning products. In DISH, Texas, elevated levels of disulphides, benzene, xylenes and naphthalene have been detected in the air, alongside numerous local complaints of headaches, diarrhoea, nosebleeds, dizziness, muscle spasms and other problems. Epidemiological studies that might confirm or rule out any connection between these complaints and fracking are virtually non-existent. Individuals "smell things that don't make them feel well, but we know nothing about cause-and-effect relationships in these cases."

In Garfield County, Colorado, another area with a high concentration of drilling rigs, volatile organic compound emissions increased 30% between 2004 and 2006; during the same period there was a rash of health complaints from local residents. The health effects of VOCs are largely unquantified, so any causal relationship is difficult to ascertain; however, some of these chemicals are suspected carcinogens and neurotoxins. Investigators from the Colorado School of Public Health performed a study in Garfield regarding potential adverse health effects, and concluded that residents near gas wells might suffer chemical exposures, accidents from industry operations, and psychological impacts such as depression, anxiety and stress. This study (the only one of its kind to date) was never published, owing to disagreements between community members and the drilling company over the study's methods. In 2010 the film *Gasland* premiered at the Sundance Film Festival. The filmmaker claims that chemicals including toxins, known carcinogens, and heavy metals polluted the ground water near well sites in Pennsylvania, Wyoming, and Colorado.

A 2011 report by the Massachusetts Institute of Technology addressed groundwater contamination, noting "There has been concern that these fractures can also penetrate shallow freshwater zones and contaminate them with fracturing fluid, but there is no evidence that this is occurring". This study blames known instances of methane contamination on a small number of sub-standard operations, and encourages the use of industry best practices to prevent such events from recurring.

Directed by Congress, the U.S. EPA announced in March 2010 that it will examine claims of water pollution related to hydraulic fracturing.

Chemicals used in Fracturing Fluid

Chemical additives used in fracturing fluids typically make up less than 1% by weight of the total fluid. They are biocides, surfactants, adjusting viscosity, and emulsifiers. Many are used in household products such as cosmetics, lotions, soaps, detergents, furniture polishes, floor waxes, and paints. Some are also used in food products. A list of the chemicals that have been used was published in a U.S. House of Representatives Report. Some of the chemicals pose no known health hazards, some of the chemicals are known carcinogens, some are toxic, and the health hazards for the rest are unknown. The report does not include the concentration of each chemical used or the amount used.

A 2011 study identified 632 chemicals used in natural gas operations. Only 353 of these are well-described in the scientific literature; and of these, more than 75% could affect skin, eyes, respiratory and gastrointestinal systems; roughly 40-50% could affect the brain and nervous, immune and cardiovascular systems and the kidneys; 37% could affect the endocrine system; and 25% were carcinogens or mutagens. The study indicated possible long-term health effects that might not appear immediately. The study recommended full disclosure of all products used, along with extensive air and water monitoring near natural gas operations; it also recommended that fracking's exemption from regulation under the US Safe Drinking Water Act be rescinded.

Earthquakes

The United States Geological Survey (USGS) reports that "Earthquakes induced by human activity have been documented in a few locations" in the United States, Japan, and Canada; "the cause was injection of fluids into deep wells for waste disposal and secondary recovery of oil, and the use of reservoirs for water supplies." The disposal and injection wells referenced are regulated under the Safe Drinking

Water Act and UIC laws and are not wells where hydraulic fracturing is generally performed.

Greenhouse Gas Emissions

The use of natural gas rather than oil or coal is sometimes touted as a way of alleviating global warming: natural gas burns more cleanly, and gas power stations can produce up to 50% less greenhouse gases than coal stations. However, an analysis of the well-to-consumer lifecycle of fracked natural gas concluded that 3.6–7.9% of the methane produced by a well will be leaked into the atmosphere during the well's lifetime. Because methane is such a potent greenhouse gas, this means that over short timescales, shale gas is actually worse than coal or oil. Methane gradually breaks down in the atmosphere, so that over long periods it is not as problematic as carbon dioxide; yet even if shale gas is burnt in efficient gas power stations, its greenhouse-gas footprint is still worse than coal or oil for timescales of less than fifty years.

Hydraulic Fracturing by Country

Hydraulic fracturing has become a contentious environmental and health issue with France banning the practice and a moratorium in place in New South Wales (Australia), Karoo basin (South Africa), Quebec (Canada), and some of the states of the US.

Australia

On 21 February 2011, the ABC's investigative journalism program Four Corners aired a program showing incidents of gas leaks into the water basin and evidence of contamination by hydraulic fracturing in Chinchilla, Queensland as a result of drilling carried out by a Queensland gas company, QGC. There is currently a moratorium in place on the practice of hydraulic fracturing in the state of New South Wales.

Canada

Fracking has been in use in Canada at an industrial level since the 1990s. Concerns about fracking began, in late July 2011, when the Government of British Columbia gave Talisman Energy a long-term water licence to draw water from the BC Hydro-owned Williston Lake reservoir, for a twenty year term. Fracking has also received criticism in New Brunswick and Nova Scotia, and the Nova Scotia government is currently reviewing the practice, with recomendations expected in March of 2012. The practice has been temporarily suspended, in Quebec, pending an environmental review. The Canadian Centre for Policy Alternatives has also expressed concern.

France

Hydraulic fracturing was banned in France in 2011 after public pressure.

New Zealand

In New Zealand, hydraulic fracturing is part of petroleum exploration and extraction on a small scale mainly in Taranaki and concerns have been raised by environmentalists.

South Africa

There is currently a moratorium on hydraulic fracturing in South Africa's Karoo region despite the interest of several energy companies.

United Kingdom

Some hydraulic fracturing is currently proceeding in the United Kingdom operated by Cuadrilla and a number of other companies. In Lancashire, operations were suspended after two small earthquakes subsequent to drilling operations. Several protest groups have started to oppose hydraulic fracturing in the UK such as nationwide groups like Frack Off and local groups like Ribble Estuary Against Fracking and The Vale Says No.

Hydraulic Fracturing in the United States

Considerable controversy surrounds the current implementation of hydraulic fracturing technology in the United States.

Hydraulic fracturing, or *fracking*, is the process of utilizing pressurized water, or some other liquid, to fracture rock layers and release petroleum, natural gas, or other substances so that they can be extracted. Environmental safety and health concerns about this process have emerged and are being debated at the state and national levels.

History

Hydraulic fracturing for stimulation of oil and natural gas wells was first used in the United States in 1947. It was first used commercially by Halliburton in 1949, and because of its success in increasing production from oil wells was quickly adopted, and is now used worldwide in tens of thousands of oil and natural gas wells annually. The first industrial use of hydraulic fracturing was as early as 1903, according to T.L. Watson. Before that date, hydraulic fracturing was used at Mt. Airy Quarry, near Mt Airy, North Carolina where it was (and still is) used to separate granite blocks from bedrock.

Volcanic dikes and sills are examples of natural hydraulic fractures. Hydraulic fracturing incorporates results from the disciplines of fracture mechanics, fluid mechanics, solid mechanics, and porous medium flow.

Contemporary Use

Hydraulic fracturing is used in mining operations and to increase the production of oil and natural gas wells. It is also used in remediating subsurface contaminants.

Environmental and Health Effects

Water and Health

In April 2010 the state of Pennsylvania banned Cabot Oil & Gas Corp. from further drilling in the entire state until it plugs wells believed to be the source of contamination of the drinking water of 14 homes in Dimock Township, Pennsylvania. The investigation was initiated after a water well exploded on New Year's Day in 2009. The state investigation revealed that Cabot Oil & Gas Company "had allowed combustible gas to escape into the region's groundwater supplies."

In the United States, a 2004 Environmental Protection Agency (EPA) study concluded that the process was safe and didn't warrant further study, because there was "no unequivocal evidence" of health risks, and the fluids were neither necessarily hazardous nor able to travel far underground. That study, however, was not intended as a general study of hydraulic fracturing, but only of its use in coalbed methane deposits, and the study did not consider impacts above ground. The EPA report did find uncertainties in knowledge of how fracturing fluid migrates through rocks, and upon its release service companies voluntarily agreed to stop using diesel fuel as a component of fracturing fluid in coalbed methane walls due to public concerns of its potential as a source of benzene contamination. Environmental group Riverkeeper presented a report to the EPA of over 100 cases of contamination. It has published a report of various environmental impacts using reports from federal and state regulators.

The increased use of hydraulic fracturing has prompted more speculation about its environmental dangers. A 2008 investigation of benzene contamination in Colorado and Wyoming led some EPA officials to suggest hydraulic fracturing as a culprit. One of the authors of the 2004 EPA report states that it has been misconstrued by the gas-drilling industry.

A 2011 study by Congressional Democrats found that, in the process of hydraulic fracturing, "oil and gas companies injected hundreds of millions of gallons of hazardous or carcinogenic chemicals into wells in more than 13 states from 2005 to 2009," according to the New York Times. A 2011 investigation by the New York Times based on various leaked EPA documents found that hydraulic fracturing had resulted in significant increases of radioactive material including radium and carcinogens including benzene in major rivers and watersheds. At one site the amount of benzene discharged into the Allegheny River after treatment was 28 times accepted levels for drinking water.

A 2011 peer-reviewed study found, on average, methane concentrations 17 times above normal in samples taken from water wells near shale gas drilling sites employing hydraulic fracturing. Water samples from 68 private water wells in the states of Pennsylvania and New York were tested and some were found to have extremely high concentrations of methane: 64 milligrams of methane per liter of drinking water, compared with a normal level of one milligram or lower. According to one of the authors of the study, "That sort of concentration is up at a level where people worry about an explosion hazard."

The average concentration of methane in the water wells near drilling sites lies within a range that, according to the U.S. Department of the Interior, is dangerous and requires urgent "hazard mitigation" action. The research was conducted by scientists at Duke University and what they found was that "levels of flammable methane gas in drinking water wells increased to dangerous levels when those water supplies were close to natural gas wells. They also found that the type of gas detected at high levels in the water was the same type of gas that energy companies were extracting from thousands of feet underground, strongly implying that the gas may be seeping underground through natural or manmade faults and fractures, or coming from cracks in the well structure itself ".

Methane contamination has been a common complaint among people who live near natural gas drilling areas. In 2009, a Propublica investigation revealed that methane contamination is widespread, "methane related to the natural gas industry has contaminated water wells in at least seven Pennsylvania counties since 2004". Because of this contamination, several homes have blown up after gas seeped into their water supplies; there have been reports of house explosions in Pennsylvania and Ohio. In one case in 2004, a methane leak caused an explosion that killed a couple and their 17 month old grandson.

A study published in May 2011 concluded that fracking has seriously contaminated shallow groundwater supplies in northeast Pennsylvania with flammable methane. However the study does not discuss how pervasive such contamination might be in other areas where drilling for shale gas has taken place.

Impact on Drinking Water Supplies

As also noted in the general article on Hydraulic Fracturing, there are also concerns about hydraulic fracturing's impact on drinking water supplies. The procedure uses millions of gallons of fresh water per well. Some states, such as Texas, Florida, Arizona, and California are trying to get ahead of drinking water shortages by recycling sewage (also called effluent, gray, and reclaimed water).

This form of processing cannot, however, remove many of the toxins used in hydraulic fracturing or all pathogens present in wastewater. Its effectiveness and safety in converting wastewater into drinking water is being debated. This technology has been used in the US space program for years.

Well Blowouts and Spills of Fracturing Fluids

A well blowout in Clearfield County, Pennsylvania on June 3, 2010, sent more than 35,000 gallons of hydraulic fracturing fluids into the air and onto the surrounding landscape in a forested area. Campers were evacuated and the company EOG Resources and the well completion company C.C. Forbes have been ordered to cease all operations in the state of Pennsylvania pending investigation. The Pennsylvania Department of Environmental Protection has called this a "serious incident".

Natural Gas Drilling and Seismic Events

Injection of fluid into subsurface geological structures, such as faults and fractures, reduces the effective normal stress acting across these structures.

If sufficient shear stress is present, the structure may slip in shear and generate seismic events over a range of magnitudes; it is possible that natural gas drilling may have caused earthquakes in North Texas. Subsidence is not directly caused by hydraulic fracturing but may occur after considerable production of oil or ground water. Subsidence occurs over reservoirs whether they have been subject to hydraulic fracturing or not because it is a result of producing fluids from the reservoir and lowering the reservoir pore pressure. The subsidence process can be associated with some seismicity. Reports of minor tremors of no greater than 2.8 on the Richter scale were reported on June 2, 2009 in Cleburne, Texas, the first in the town's 140-year history.

Fracking connected to deep well waste water disposal has been banned in Arkansas due to the risk of contaminating aquifers if the earthquakes it induced continued.

Air and Health

A potential hazard that is commonly overlooked is the venting of bulk sand silos directly to atmosphere. When they are being filled, or emptied during the fracture, a fine cloud of silica particulate will be vented directly into atmosphere. This dust has the potential to travel many kilometres on the wind directly into populated areas.

While the immediate personnel are wearing personal protective equipment, other people in the area of a well fracture can potentially be exposed. Many particulates and chemicals can be released into the atmosphere, such as sulfuric Oxide, nitrous oxides, volatile organic compounds (VOC), benzene, toluene, diesel fuel, hydrogen sulfide which can have serious health implications.

Other Consequences

A 2011 Cornell University study found that natural gas extracted by hydraulic fracturing may contribute as much to global warming as coal, or more so. The authors of the study acknowledge, however, that the data they used were not the best available, so this has left the study open to interpretation. The natural gas industry has noted significant problems with the study's methodology, including comments from the authors that acknowledge the data may be unreliable.

There are also potential community complications as a result of fracking. When drilling companies move into a new area the population increases and with it comes problems related to population boom. There is the potential for noise and light pollution complaints, reports of crime can go up, motor vehicle accidents increase, traffic and road degradation increase, sexually transmitted infections increase, and strain on schools are all some potential problems facing communities where gas drilling is nearby.

In Garfield county, Colorado the Colorado School of Public Health released a second draft report released the Battlement Mesa Health Impact Assessment on March 1, 2011 for public comment.

University of Michigan-Ann Arbor in May 2011 began studying the effects of natural gas fracking on the Great Lakes.

Congress has requested that the EPA undertake a new, broader study of hydraulic fracturing. The report is due to be released in 2012.

Occupational Hazards

The EPA states in their Hydraulic Fracturing Study Plan (2011) that the exposure to hydraulic fracturing chemicals in an occupational setting needs to be examined to determine the acute and chronic effects on health. The exposure risks such as "transport, mixing, delivery, and potential accidents" have not been properly assessed.

Hydraulic fracturing also affects individuals in close proximity to the activity, like the case previously discussed about the nurse who became ill after exposure from treating a hydraulic fracturing worker (Frankowski, 2008).

Economic Impacts

The economic impacts of hydraulic fracturing include property owners that receive funds, an increase in jobs, and an increase in business. The EPA states that it is unclear on a local level how and for how long hydraulic fracturing affects a community economically. It is hypothesized that hydraulic fracturing may not provide jobs to local communities due to the specialized nature of hydraulic fracturing tasks. Also, communities' local resources could potentially be taxed due to the increase in industry traffic or if an accident occurs.

Furthermore, a June, 2011 New York Times investigation of industrial emails and internal documents found that the financial benefits of unconventional shale gas extraction may be less than previously thought, due to companies intentionally overstating the productivity of their wells and the size of their reserves.

Lawsuits

In September 2010, a lawsuit was filed in Pennsylvania alleging that Southwestern Energy Company contaminated aquifers through a defective cement casing in the well.

In May 2011, Jessica Ernst launched a multi million dollar lawsuit against Encana Corporation, the Alberta Energy Resources Conservation Board, and the Alberta government for contamination of her property and drinking water due to Encana's fracking program. Encana fractures rock to extract coal bed methane, much as fracking is used to extract natural gas from shale. (In March, after a public hearing, Quebec put a moratorium on shale gas exploration pending a full environmental assessment of the potential damage from fracking.) According to the

Statement of Claim, many Albertans depend on drinking water from coal bed aquifers, but Ms.

Ernst's water is now so contaminated that it can be lit on fire. She is also suing Alberta's oil and gas regulator, alleging that it not only tolerated illegal behaviour by Encana and failed to protect her, but actively attempted to silence her complaints, and that Alberta Environment showed bad faith in "investigating" those complaints. The lawsuit, together with the Quebec moratorium, signals the likelihood of stronger environmental regulations of fracking in the pursuit of shale gas or coalbed methane.

In June 2011, Northeast Natural Energy sued the town of Morgantown, West Virginia, for its ban on hydraulic fracturing in the Marcellus Shale within a mile of the town's borders. The ban had been initiated to protect the municipal water supply as well as the town's inhabitants, in the absence at the state level of regulations specific to hydraulic fracturing.

Regulation

On August 7, 1997, the Eleventh Circuit Court ordered the United States Environmental Protection Agency to reevaluate its stance on hydraulic fracturing based on a lawsuit brought by the Legal Environmental Assistance Foundation.

Up until that decision, the EPA deemed that hydraulic fracturing did not fall under the rules in the Safe Drinking Water Act.

While the impact of this decision was localized to Alabama, it forced the EPA to evaluate its oversight responsibility under the Safe Drinking Water Act for hydraulic fracturing. In 2004, the EPA released a study that concluded the threat to drinking water from hydraulic fracturing was "minimal" and the Energy Policy Act of 2005 exempted fractured wells from being re-classified as injection wells, which would place them under federal regulation under the Safe Drinking Water Act. which was originally intended to regulate disposal wells.

In April 2011, the Ground Water Protection Council, in conjunction with the industry, began releasing well-by-well lists of hydraulic fracturing chemicals at.

A complete listing of the specific chemical formulation of additives used in hydraulic fracturing operations is not currently made available to landowners, neighbors, local officials, or health care providers. This practice is under scrutiny as well.

Two studies released in 2009, one by the U.S. Department of Energy and the other released by the Ground Water Protection Council, address hydraulic fracturing safety concerns. Chemicals which can be used in the fracturing fluid include kerosene, benzene, toluene, xylene, and formaldehyde. These chemicals are not directly used as treating chemical additives but can be a small component of the specific chemicals used in the job.

On June 8, 2010 the Wyoming Oil and Gas Conservation Commission voted to require full disclosure of the hydraulic fracturing fluids used in natural gas exploration. This may aid in tracking pollutants that have migrated from hydraulically fractured gas wells.

Congress has been urged to repeal the 2005 regulatory exemption under the Energy Policy Act of 2005. The FRAC Act, introduced in June 2009, would eliminate the exemption and might allow producing wells to be reclassified as injection wells placing them under federal jurisdiction in states without approved UIC programs.

The New York City watershed includes a large area of the Marcellus shale formation. The NYC Dept. of Environmental Protection's position: "While DEP is mindful of the potential economic opportunity that this represents for the State, hydraulic fracturing poses an unacceptable threat to the unfiltered water supply of nine million New Yorkers and cannot safely be permitted with the New York City watershed."

The New York State assembly voted 93 to 43, Nov. 30, 2010, to place a moratorium or freeze on hydraulic fracturing to give the state more time to undertake safety and environmental concerns.

At the municipal level, some towns and cities in central New York state have moved to regulate drilling by hydraulic fracturing and its attendant effects, either by banning it within municipal limits, maintaining the option to do so in the future, or banning wastewater from the drilling process from municipal water treatment plants.

EPA Hydraulic Fracturing Study

The purpose of the EPA study regarding Hydraulic Fracturing is to examine the effects of hydraulic fracturing on the water supply, specifically for human consumption. The research aims to examine the full scope of the water pathway as it moves through the hydraulic fracturing process, including water that is used for the construction of the wells, the fracturing mixture, and subsequent removal and disposal. Fundamental research questions include:

- How might large volume water withdrawals from ground and surface water impact drinking water resources?
- What are the possible impacts of releases of flowback and produced water on drinking water resources?
- What are the possible impacts of inadequate treatment of hydraulic fracturing wastewaters on drinking water resources?

The Scientific Advisory Board reviewed the study plan in early March 2011. In June 2011, the EPA announced the locations of its five retrospective case studies, which will examine existing hydrofracking sites for evidence of drinking water contamination. They are:

- Bakken Shale - Kildeer, and Dunn Counties, North Dakota
- Barnett Shale - Wise and Denton Counties, Texas
- Marcellus Shale - Bradford and Susquehanna Counties, Pennsylvania
- Marcellus Shale - Washington County, Pennsylvania
- Raton Basin - Las Animas County, Colorado.

Research should be completed by the end of 2012, and the EPA's Hydraulic Fracturing Report is expected to be completed in 2014.

Fracturing Responsibility and Awareness of Chemicals Act

In June 2009 two identical bills named the Fracturing Responsibility and Awareness of Chemicals Act were introduced to both the United States House and the Senate. The House bill was introduced by representatives Diana DeGette, D-Colo., Maurice Hinchey D-N.Y., and Jared Polis, D-Colo.

The Senate version was introduced by senators Bob Casey, D-Pa., and Chuck Schumer, D-N.Y. These bills are designed to amend the Safe Drinking Water Act. This would allow the Environmental Protection Agency to regulate hydraulic fracturing that occurs in states which have not taken primacy in UIC regulation. The bill required the energy industry to reveal what chemicals are being used in the sand-water mixture. The 111th Congress adjourned (Jan. 3, 2011) without taking action on the FRAC Act. The 112th Congress has not re-introduced the bill or an equivalent.

Halliburton's Hydraulic Fracturing Operations

As mandated by EPA subpoena, Halliburton has released hydraulic fracturing operations. Details regarding Halliburton's fracturing process and fracturing chemicals can be found on their website.

Fracturing Method: High Pressure by Combustible Gas Mixtures or Driving Forces

Fracturing is done by pumping in liquids at high pressure. In the past, combustible gas mixtures, driving liquids or explosives generated high-pressure high-speed gas flow (TNT or PETN up to 1,900,000 psi). In the late 1960s and early 1970s, as part of Operation Plowshare, underground nuclear explosions were tested for natural gas stimulation. The Rulison explosion multiplied the accessibility of the gas, but the gas was contaminated and unmarketable. Stimulation by nuclear explosion is not currently utilized by the industry.

List of Additives for Hydraulic Fracturing

About 750 chemicals have been listed as additives for hydraulic fracturing in a report to the US Congress in 2011. The following is a partial list of the chemical constituents in additives used in fracturing operations, as indicated by the New York State Department of Environmental Conservation. Each state has a contact person in charge of the fracking regulations of these additives.

CAS Number	*Chemical Constituent*
2634-33-5	1,2-Benzisothiazolin-2-one/1,2-benzisothiazolin-3-one
95-63-6	1,2,4-trimethylbenzene
123-91-1	1,4-Dioxane
3452-07-1	1-eicosene
629-73-2	1-hexadecene
112-88-9	1-octadecene
1120-36-1	1-tetradecene
10222-01-2	2,2 Dibromo-3-nitrilopropionamide, a biocide
27776-21-2	2,2'-azobis-{2-(imidazlin-2-yl)propane}-dihydrochloride
73003-80-2	2,2-Dibromomalonamide
15214-89-8	2-Acrylamido-2-methylpropane sulphonic acid sodium salt polymer
46830-22-2	2-acryloyloxyethyl(benzyl)dimethylammonium chloride
52-51-7	2-Bromo-2-nitro-1,3-propanediol
111-76-2	2-Butoxy ethanol
1113-55-9	2-Dibromo-3-Nitriloprionamide (2-Monobromo-3-nitriilopropionamide)
104-76-7	2-Ethyl Hexanol
67-63-0	2-Propanol/Isopropyl Alcohol/Isopropanol/Propan-2-ol

Contd...

CAS Number	***Chemical Constituent***
26062-79-3	2-Propen-1-aminium, N,N-dimethyl-N-2-propenyl-chloride, homopolymer
9003-03-6	2-propenoic acid, homopolymer, ammonium salt
25987-30-8	2-Propenoic acid, polymer with 2 p-propenamide, sodium salt/Copolymer of acrylamide and sodium acrylate
71050-62-9	2-Propenoic acid, polymer with sodium phosphinate (1:1)
66019-18-9	2-propenoic acid, telomer with sodium hydrogen sulfite
107-19-7	2-Propyn-1-ol/Propargyl alcohol
51229-78-8	3,5,7-Triaza-1-azoniatricyclo[3.3.1.13,7]decane, 1-(3-chloro-2-propenyl)-chloride,
115-19-5	3-methyl-1-butyn-3-ol
127087-87-0	4-Nonylphenol Polyethylene Glycol Ether Branched/ Nonylphenol ethoxylated/Oxyalkylated Phenol
64-19-7	Acetic acid
68442-62-6	Acetic acid, hydroxy-, reaction products with triethanolamine
108-24-7	Acetic Anhydride
67-64-1	Acetone
79-06-1	Acrylamide
38193-60-1	Acrylamide - sodium 2-acrylamido-2-methylpropane sulfonate copolymer
25085-02-3	Acrylamide - Sodium Acrylate Copolymer or Anionic Polyacrylamide
69418-26-4	Acrylamide polymer with N,N,N-trimethyl-2[1-oxo-2-propenyl]oxy Ethanaminium chloride
15085	-02-3 Acrylamide-sodium acrylate copolymer
68551-12-2	Alcohols, C12-C16, Ethoxylated (a.k.a. Ethoxylated alcohol)
64742-47-8	Aliphatic Hydrocarbon/Hydrotreated light distillate/ Petroleum Distillates/Isoparaffinic Solvent/Paraffin Solvent/Napthenic Solvent
64743-02-8	Alkenes
68439-57-6	Alkyl (C14-C16) olefin sulfonate, sodium salt
9016-45-9	Alkylphenol ethoxylate surfactants
1327-41-9	Aluminum chloride
73138-27-9	Amines, C12-14-tert-alkyl, ethoxylated
71011-04-6	Amines, Ditallow alkyl, ethoxylated
68551-33-7	Amines, tallow alkyl, ethoxylated, acetates

Contd...

CAS Number	***Chemical Constituent***
1336-21-6	Ammonia
631-61-8	Ammonium acetate
68037-05-8	Ammonium Alcohol Ether Sulfate
7783-20-2	Ammonium bisulfate
10192-30-0	Ammonium bisulfite
12125-02-9	Ammonium chloride
7632-50-0	Ammonium citrate
37475-88-0	Ammonium Cumene Sulfonate
1341-49-7	Ammonium hydrogen-difluoride
6484-52-2	Ammonium nitrate
7727-54-0	Ammonium Persulfate/Diammonium peroxidisulphate
1762-95-4	Ammonium Thiocyanate
7664-41-7	Aqueous ammonia
121888-68-4	Bentonite, benzyl(hydrogenated tallow alkyl) dimethylammonium stearate complex/organophilic clay
71-43-2	Benzene
119345-04-9	Benzene, 1,1'-oxybis, tetratpropylene derivatives, sulfonated, sodium salts
74153-51-8	Benzenemethanaminium, N,N-dimethyl-N-[2-[(1-oxo-2-propenyl)oxy]ethyl]-, chloride, polymer with 2-propenamide
10043-35-3	Boric acid
1303-86-2	Boric oxide/Boric Anhydride
71-36-3	Butan-1-ol
68002-97-1	C10 - C16 Ethoxylated Alcohol
68131-39-5	C12-15 Alcohol, Ethoxylated
10043-52-4	Calcium chloride
124-38-9	Carbon dioxide
68130-15-4	Carboxymethylhydroxypropyl guar
9012-54-8	Cellulase/Hemicellulase Enzyme
9004-34-6	Cellulose
10049-04-4	Chlorine dioxide
77-92-9	Citric Acid
94266-47-4	Citrus Terpenes
61789-40-0	Cocamidopropyl betaine
68155-09-9	Cocamidopropylamine Oxide
68424-94-2	Coco-betaine

Contd...

CAS Number	***Chemical Constituent***
7758-98-7	Copper(II) sulfate
31726-34-8	Crissanol A-55
14808-60-7	Crystalline Silica (Quartz)
7447-39-4	Cupric chloride dihydrate
1120-24-7	Decyldimethyl Amine
2605-79-0	Decyl-dimethyl Amine Oxide
3252-43-5	Dibromoacetonitrile
25340-17-4	Diethylbenzene
111-46-6	Diethylene glycol
22042-96-2	Diethylenetriamine penta (methylenephonic acid) sodium salt
28757-00-8	Diisopropyl naphthalenesulfonic acid
68607-28-3	Dimethylcocoamine, bis(chloroethyl) ether, diquaternary ammonium salt
7398-69-8	Dimethyldiallylammonium chloride
25265-71-8	Dipropylene glycol
139-33-3	Disodium Ethylene Diamine Tetra Acetate
5989-27-5	D-Limonene
123-01-3	Dodecylbenzene
27176-87-0	Dodecylbenzene sulfonic acid
42504-46-1	Dodecylbenzenesulfonate isopropanolamine
50-70-4	D-Sorbitol/Sorbitol
37288-54-3	Endo-1,4-beta-mannanase, or Hemicellulase
149879-98-1	Erucic Amidopropyl Dimethyl Betaine
89-65-6	Erythorbic acid, anhydrous
54076-97-0	Ethanaminium, N,N,N-trimethyl-2-[(1-oxo-2 propenyl)oxy]-, chloride, homopolymer
107-21	-1 Ethane-1,2-diol/Ethylene Glycol
9002-93-1	Ethoxylated 4-tert-octylphenol
68439-50-9	Ethoxylated alcohol
126950-60-5	Ethoxylated alcohol
67254-71-1	Ethoxylated alcohol (C10-12)
68951-67-7	Ethoxylated alcohol (C14-15)
68439-46-3	Ethoxylated alcohol (C9-11)
66455-15-0	Ethoxylated Alcohols
84133-50-6	Ethoxylated Alcohols (C12-14 Secondary)

Contd...

CAS Number	***Chemical Constituent***
68439-51-0	Ethoxylated Alcohols (C12-14)
78330-21-9	Ethoxylated branch alcohol
34398-01-1	Ethoxylated C11 alcohol
61791-12-6	Ethoxylated Castor Oil
61791-29-5	Ethoxylated fatty acid, coco
61791-08-0	Ethoxylated fatty acid, coco, reaction product with ethanolamine
68439-45-2	Ethoxylated hexanol
9036-19-5	Ethoxylated octylphenol
9005-67-8	Ethoxylated Sorbitan Monostearate
9004-70-3	Ethoxylated Sorbitan Trioleate
64-17-5	Ethyl alcohol/ethanol
100-41-4	Ethyl Benzene
97-64-3	Ethyl lactate
9003-11-6	Ethylene Glycol-Propylene Glycol Copolymer (Oxirane, methyl-, polymer with oxirane)
75-21-8	Ethylene oxide
5877-42-9	Ethyloctynol
68526-86-3	Exxal 13
61790-12-3	Fatty Acids
68188-40-9	Fatty acids, tall oil reaction products w/acetophenone, formaldehyde & thiourea
9043-30-5	Fatty alcohol polyglycol ether surfactant
7705-08-0	Ferric chloride
7782-63-0	Ferrous sulfate, heptahydrate
50-00-0	Formaldehyde
29316-47-0	Formaldehyde polymer with 4,1,1-dimethylethyl phenolmethyl oxirane
153795-76-7	Formaldehyde, polymers with branched 4-nonylphenol, ethylene oxide and propylene oxide
75-12-7	Formamide
64-18-6	Formic acid
110-17-8	Fumaric acid
65997-17-3	Glassy calcium magnesium phosphate
111-30-8	Glutaraldehyde
56-81-5	Glycerol/glycerine

Contd...

CAS Number	*Chemical Constituent*
9000-30-0	Guar Gum
64742-94-5	Heavy aromatic petroleum naphtha
9025-56-3	Hemicellulase
7647-01-0	Hydrochloric Acid/Hydrogen Chloride/muriatic acid
7722-84-1	Hydrogen peroxide
79-14-1	Hydroxy acetic acid
35249-89-9	Hydroxyacetic acid ammonium salt
9004-62-0	Hydroxyethyl cellulose
5470-11-1	Hydroxylamine hydrochloride
39421-75-5	Hydroxypropyl guar
35674-56-7	Isomeric Aromatic Ammonium Salt
64742-88-7	Isoparaffinic Petroleum Hydrocarbons, Synthetic
64-63-0	Isopropanol
98-82-8	Isopropylbenzene (cumene)
68909-80-8	Isoquinoline, reaction products with benzyl chloride and quinoline
8008-20-6	Kerosene
64742-81-0	Kerosine, hydrodesulfurized
63-42-3	Lactose
64742-95-6	Light aromatic solvent naphtha
1120-21-4	Light Paraffin Oil
14807-96-6	Magnesium Silicate Hydrate (Talc)
1184-78-7	methanamine, N,N-dimethyl-, N-oxide
67-56-1	Methanol
68891-11-2	Methyloxirane polymer with oxirane, mono (nonylphenol) ether, branched
8052-41-3	Mineral spirits/Stoddard Solvent
141-43-5	Monoethanolamine
44992-01-0	N,N,N-trimethyl-2[1-oxo-2-propenyl]oxy Ethanaminium chloride
64742-48-9	Naphtha (petroleum), hydrotreated heavy
91-20-3	Naphthalene
38640-62-9	Naphthalene bis(1-methylethyl)
93-18-5	Naphthalene, 2-ethoxy-
68909-18-2	N-benzyl-alkyl-pyridinium chloride

Contd...

CAS Number	***Chemical Constituent***
68139-30-0	N-Cocoamidopropyl-N,N-dimethyl-N-2-hydroxypropylsulfobetaine
7727-37-9	Nitrogen, Liquid form
68412-54-4	Nonylphenol Polyethoxylate
121888-66-2	Organophilic Clays
64742-65-0	Petroleum Base Oil
64741-68-0	Petroleum naphtha
70714-66-8	Phosphonic acid, [[(phosphonomethyl)imino]bis[2,1-ethanediylnitrilobis(methylene)]]tetrakis-, ammonium salt
8000-41-7	Pine Oil
60828-78-6	Poly(oxy-1,2-ethanediyl), a-[3,5-dimethyl-1-(2-methylpropyl)hexyl]-w-hydroxy-
25322-68-3	Poly(oxy-1,2-ethanediyl), a-hydro-w-hydroxy/Polyethylene Glycol
24938-91-8	Poly(oxy-1,2-ethanediyl), α-tridecyl-ù-hydroxy-
51838-31-4	Polyepichlorohydrin, trimethylamine quaternized
56449-46-8	Polyethlene glycol oleate ester
62649-23-4	Polymer with 2-propenoic acid and sodium 2-propenoate
9005-65-6	Polyoxyethylene Sorbitan Monooleate
61791-26-2	Polyoxylated fatty amine salt
127-08-2	Potassium acetate
12712-38-8	Potassium borate
1332-77-0	Potassium borate
20786-60-1	Potassium Borate
584-08-7	Potassium carbonate
7447-40-7	Potassium chloride
590-29-4	Potassium formate
1310-58-3	Potassium Hydroxide
13709-94-9	Potassium metaborate
24634-61-5	Potassium sorbate
112926-00-8	Precipitated silica/silica gel
57-55-6	Propane-1,2-diol, or Propylene glycol
107-98-2	Propylene glycol monomethyl ether
68953-58-2	Quaternary Ammonium Compounds
62763-89-7	Quinoline,2-methyl-, hydrochloride

Contd...

CAS Number	***Chemical Constituent***
15619-48-4	Quinolinium, 1-(phenylmethl),chloride
7631-86-9	Silica, Dissolved
5324-84-5	Sodium 1-octanesulfonate
127-09-3	Sodium acetate
95371-16-7	Sodium Alpha-olefin Sulfonate
532-32-1	Sodium benzoate
144-55-8	Sodium bicarbonate
7631-90-5	Sodium bisulfate
7647-15-6	Sodium bromide
497-19-8	Sodium carbonate
7647-14-5	Sodium Chloride
7758-19-2	Sodium chlorite
3926-62-3	Sodium chloroacetate
68-04-2	Sodium citrate
6381-77-7	Sodium erythorbate/isoascorbic acid, sodium salt
2836-32-0	Sodium Glycolate
1310-73-2	Sodium Hydroxide
7681-52-9	Sodium hypochlorite
7775-19-1	Sodium Metaborate .8H2O
10486-00-7	Sodium perborate tetrahydrate
7775-27-1	Sodium persulfate
9003-04-7	Sodium polyacrylate
7757-82-6	Sodium sulfate
1303-96-4	Sodium tetraborate decahydrate
7772-98-7	Sodium thiosulfate
1338-43-8	Sorbitan Monooleate
57-50-1	Sucrose
5329-14-6	Sulfamic acid
112945-52-5	Synthetic Amorphous/Pyrogenic Silica/Amorphous Silica
68155-20-4	Tall Oil Fatty Acid Diethanolamine
8052-48-0	Tallow fatty acids sodium salt
72480-70-7	Tar bases, quinoline derivs., benzyl chloride-quaternized
68647-72-3	Terpene and terpenoids
68956-56-9	Terpene hydrocarbon byproducts

Contd...

CAS Number	*Chemical Constituent*
533-74-4	Tetrahydro-3,5-dimethyl-2H-1,3,5-thiadiazine-2-thione (a.k.a. Dazomet)
55566-30-8	Tetrakis(hydroxymethyl)phosphonium sulfate (THPS)
75-57-0	Tetramethyl ammonium chloride
64-02-8	Tetrasodium Ethylenediaminetetraacetate
68-11-1	Thioglycolic acid
62-56-6	Thiourea
68527-49-1	Thiourea, polymer with formaldehyde and 1-phenylethanone
108-88-3	Toluene
81741-28-8	Tributyl tetradecyl phosphonium chloride
68299-02-5	Triethanolamine hydroxyacetate
112-27-6	Triethylene glycol
52624-57-4	Trimethylolpropane, Ethoxylated, Propoxylated
150-38-9	Trisodium Ethylenediaminetetraacetate
5064-31-3	Trisodium Nitrilotriacetate
7601-54-9	Trisodium orthophosphate
57-13-6	Urea
25038-72-6	Vinylidene Chloride/Methylacrylate Copolymer
7732-18-5	Water
1330-20-7	Xylene
	Aliphatic acids
	Aliphatic alcohol glycol ether
	Alkyl Aryl Polyethoxy Ethanol
	Alkylaryl Sulfonate
	Aromatic hydrocarbons
	Aromatic ketones
	Oxyalkylated alkylphenol
	Petroleum distillate blend
	Polyethoxylated alkanol
	Polymeric Hydrocarbons
	Salt of amine-carbonyl condensate
	Salt of fatty acid/polyamine reaction product
	Sugar
	Surfactant blend

Shale Oil Extraction

Shale oil extraction is an industrial process for unconventional oil production. This process converts kerogen in oil shale into shale oil by pyrolysis, hydrogenation, or thermal dissolution. The resultant shale oil is used as fuel oil or upgraded to meet refinery feedstock specifications by adding hydrogen and removing sulfur and nitrogen impurities.

Shale oil extraction is usually performed above ground (*ex situ* processing) by mining the oil shale and then treating it in processing facilities. Other modern technologies perform the processing underground (on-site or *in situ* processing) by applying heat and extracting the oil via oil wells.

The earliest description of the process dates to the 10th century. In 1684, Great Britain granted the first formal extraction process patent. Extraction industries and innovations became widespread during the 19th century. The industry shrank in the mid-20th century following the discovery of large reserves of conventional oil, but high petroleum prices at the beginning of the 21st century have led to renewed interest, accompanied by the development and testing of newer technologies.

As of 2010, major long-standing extraction industries are operating in Estonia, Brazil, and China. Its economic viability usually requires a lack of locally available crude oil. National energy security issues have also played a role in its development. Critics of shale oil extraction pose questions about environmental management issues, such as waste disposal, extensive water use, waste water management, and air pollution.

History of the Oil Shale Industry

The history of the oil shale industry started in ancient times. The modern industrial use of oil shale for oil extraction dates to the mid-19th century and started growing just before World War I because of the mass production of automobiles and trucks and the supposed shortage of gasoline for transportation needs. Between the World Wars oil shale projects were begun in several countries, but after World War II, the oil shale industry was phased out in almost all countries because of high processing costs and the discovery of large supplies of easily-accessible crude oil. As of 2010, oil shale was commercially used in Estonia, China and Brazil, while several countries are considering to start or re-start commercial use of oil shale.

Early History

Humans have used oil shale as a fuel since prehistoric times, since it generally burns without any processing. It was also used for decorative purposes and construction. Britons of the Iron Age used to polish and form oil shale into ornaments. Around 3000 BC, "rock oil" was used in Mesopotamia for road construction and making architectural adhesives. As a decorative material, oil shale was also used over the Greek, Roman, Byzantinian, Umayyad and Abbasid periods to decorate mosaics and floors of the palaces, churches and mosques. Shale oil was used for medical and military purposes. Mesopotamians used it for medical purposes and for caulking ships, Mongols used to cap their arrows with flaming oil shale. In 10th century, the Arabian physician Masawaih al-Mardini (Mesue the Younger) described a method of extraction of oil from "some kind of bituminous shale".

In the early 14th century, the first use of shale oil was recorded in Switzerland and Austria. In 1350, a knight Berthold von Eberhausen was awarded a right to exploit the Seefeld oil shale in Tyrol. Oil shale was used for production of shale oil using an early retorting method of heating the crushed oil shale put in crucibles. The healing properties of a mineral oil distilled from oil shale were noted in 1596 by the personal physician of the Duke of Württemberg Frederick I. In Skåne, the Swedish alum shale dating from the Cambrian and Ordovician periods was used for extracting potassium aluminium sulfate by roasting it over fire as early as 1637.

In Italy, shale oil was used to light the streets of Modena at the turn of the 17th century. The British Crown granted a patent in 1694 to three persons who had "found a way to extract and make great quantities of pitch, tarr, and oyle out of a sort of stone." Shale oil was produced by extracting Shropshire oil shale. Later sold as Betton's British Oil, the distilled product was said to have been "tried by divers persons in Aches and Pains with much benefit. In 1781, Archibald Cochrane, 9th Earl of Dundonald, registered a patent for an extraction process to produce tar, pitch and oil from coal and bituminous shales, using masonry retorts and wooden condensers.

Start of the Modern Industry

The modern industrial use of oil shale for oil extraction started in France, where oil shale commercial mining began in Autun in 1837. The shale oil production started in 1838 by using Selligue process, invented by Alexander Selligue. In 1847 the Scottish chemist James

Young prepared "lighting oil," lubricating oil and wax from cannel coal and since 1862 from torbanite. In 1850 he patented the process of cracking oil. Commercial scale shale oil extraction in Scotland started in 1859 by Robert Bell in Broxburn, West Lothian. About the same time Germany began exploiting its deposits. During the second half of the 19th century shale oil extraction industries were initiated also in Sweden, Australia, Brazil, New Zealand, Canada and the United States. In the United States, the commercial scale shale oil extraction began in 1857 at shale oil retorts retorting the Devonian oil shale along the Ohio River Valley. In Canada, the Craigleith Shale Oil Works started to retort oil shale of the Ordovician Whitby Formation near Collingwood, Ontario, on Lake Huron, in 1859. However, after crude oil discovery in Pennsylvania in 1859, the United States and Canadian oil shale industries found it difficult to compete and they were shut down by 1861.

In Australia, the first oil shale mine was commenced in 1865 at American Creek, Mount Kembla in New South Wales. At the same year, the first shale oil was produced by the Pioneer Kerosene Works at American Creek. A number of other mines were opened in the areas of the Great Dividing Range; however, by the turn of the 20th century they were closed because of the importation of cheaper crude oil. In Brazil, oil shale was first exploited in 1884 in Bahia.

In 1894, the Pumpherston retort (also known as the Bryson retort) was invented, which is considered as a separation of the oil shale industry from the coal industry. It stayed in use until 1938.

Operations during the 19th century focused on the production of kerosene, lamp oil, and paraffin; these products helped supply the growing demand for lighting that arose during the Industrial Revolution. Fuel oil, lubricating oil and grease, and ammonium sulfate were also produced.

First Half of the 20th Century

The oil shale industry expanded immediately before World War I because of limited access to conventional petroleum resources and the mass production of automobiles and trucks, which accompanied an increase in gasoline consumption. The Office of Naval Petroleum and Oil Shale Reserves of the United States was established in 1912. The reserves were seen as a possible emergency source of fuel for the military, particularly the Navy. In 1915 an oil shale industry was established in Switzerland. The year 1916 is considered the beginning of the Estonian oil shale industry. In 1917, Russian paleontologist

Mikhail Zalessky named kukersite oil shale after the Kukruse settlement. Continuous mining activities started shortly after. Initially, oil shale was used primarily in the cement industry, for firing in locomotive furnaces, and as a household fuel, followed by shale oil and power production. The first experimental oil shale processing retorts were built in 1921. In 1924, the Tallinn Power Plant was the first power plant in the world to employ oil shale as its primary fuel.

Between the World Wars oil shale projects were begun in Spain, China, Russia and South Africa; they restarted in Brazil and, for a short time, in Canada. In China, the extraction of oil shale began in 1926 under the Japanese rule. The commercial-scale production of shale oil began in 1930 in Fushun, Manchuria, with the construction of the "Refinery No. 1" operating Fushun-type retorts. In 1939–1945, a shale oil pilot plant operated in Morocco.

During World War II a modified *in situ* extraction process was used in Germany by digging tunnels into oil shale deposits and then collapsing their walls into the void volume to ignitate. This process had extremely low oil recovery and it was hard to control. In Sweden, the Swedish Shale Oil Company was formed in 1940. It exploited one of the earliest *in-situ* processes–underground gasification by electrical energy (Ljungström method)–between 1940 and 1966 at Kvarntorp. In 1940–1952, three NTU retorts were operated at Marangaroo, New South Wales, Australia.

1950s–1960s

Although the Estonian, Russian and Chinese oil shale industries continued to grow after World War II, most other countries abandoned their projects due to high processing costs and the availability of cheaper petroleum. The shale oil extraction in Australia was discontinued in 1952 due to ceasing of government funding, in France in 1957, in Scotland and South Africa in 1962, and in Sweden and Spain in 1966. In Germany only Rohrbach Zement (now part of Holcim) in Dotternhausen continued using oil shale for cement, power and thermal energy production.

In 1946–1952, Dictyonema argillite was used for uranium production in Sillamäe. In Sweden the extraction of alum shale for uranium and vanadium continued until 1989. After World War II, Estonian-produced oil shale gas was used in Leningrad (Saint Petersburg) and in northern Estonia cities as a substitute for natural gas. In 1949, the Kohtla-Järve Power Plant – the first power plant in the world using pulverized oil shale at an industrial scale – was commissioned in Estonia. The world's two largest oil shale-fired power stations – Balti Power Plant and Eesti Power Plant (known as the Narva Power Plants) – were opened in 1965 and in 1973.

In China, the "Refinery No. 2" of Fushun began its production in 1954 and in 1959, the maximum annual shale oil production increased to 780,000 tonnes. The produced shale oil was used for producing light liquid fuels. In 1961, China was producing one third of its total oil production from oil shale.

The United States Bureau of Mines opened a demonstration mine at Anvils Point, west of Rifle, Colorado, which operated at a small scale. In the early 1960s TOSCO (The Oil Shale Corporation) opened an underground mine and built an experimental plant near Parachute, Colorado. It was closed in 1972 because the price of production exceeded the cost of imported crude oil. In 1951, the United States Department of Defence became interested in oil shale as an alternative resource for producing a jet fuel. In 1953, Sinclair Oil Corporation developed an *in situ* processing method using existing and induced fractures between vertical wells. In the 1960s, a proposal was suggested for a modified *in situ* process which involved creation of a rubble chimney (a zone in the rock formation created by breaking the rock into fragments) using a nuclear explosive.

1970s–1980s

In 1972, the first modified *in situ* oil shale experiment in the United States was conducted by Occidental Petroleum at Logan Wash, Colorado. Due to the 1973 oil crisis, the oil shale industry restarted in several countries. The United States Navy and the Office of Naval Petroleum and Oil Shale Reserves started evaluations of oil shale's suitability for military fuels, such as jet fuels, marine fuels and a heavy fuel oil. Shale-oil based JP-4 jet fuel was produced until the early 1990s, when it was replaced with kerosene-based JP-8. In 1974 the United States Department of the Interior announced an oil shale leasing program in the oil shale regions of Colorado and Utah, and by the early 1980s almost all of the major oil companies had established oil shale pilot projects.

The United States oil shale industry collapsed when oil prices fell in the early 1980s. On 2 May 1982, known as "Black Sunday", Exxon canceled its US$5 billion Colony Shale Oil Project near Parachute, Colorado because of low oil-prices and increased expenses, laying off more than 2,000 workers and leaving a trail of home-foreclosures and small-business bankruptcies. In 1986, President Ronald Reagan signed into law the Consolidated Omnibus Budget Reconciliation Act of 1985 which among other things abolished the United States' Synthetic Liquid Fuels Program. The last oil shale retort in the United States, operated by Unocal Corporation, closed in 1991.

Because of the success of oil shale-based power generation, Estonian oil shale production peaked in 1980 at 31.35 million tonnes. The largest oil shale mine in the world – the Estonia Mine – was opened in 1972. However, production decreased in Estonia during the 1990s, due to reduced demand from the power generation industry. It was mainly affected by construction of the nuclear power plants in the Soviet Union. Most of Russian oil shale mines were closed in the 1990s and production continued only on a small scale.

In Israel, a 0.1 MW pilot oil shale-fired power plant was tested in 1982–1986. A demonstration fluidised-bed plant became operational in 1989.

Latest Developments

The global oil shale industry started to grow slightly in the mid-1990s. In 1992 commercial shale oil production using Petrosix technology resumed in Brazil. Estonian oil shale production has continuously increased since 1995. In 2005, China became the largest shale oil producer in the world. In Australia, a demonstration-scale processing plant at the Stuart Deposit near Gladstone, Queensland, produced between 2000 and 2004 over 1.5 million barrels (240×10^3 m^3) of shale oil. In 2008–2009, the facility was dismantled. In the United States, an oil shale development program was initiated in 2003. The Energy Policy Act of 2005 introduced a commercial leasing program for oil shale and tar sands resources on public lands within the states of Colorado, Utah, and Wyoming.

In April 2010, the 4th Workshop on Regional Cooperation for Clean Utilization of Oil Shale was held in Egypt and later the same month an Oil Shale Cooperation Centre was established in Amman by Egypt, Jordan, Morocco, Syria and Turkey.

Processing Principles

Shale oil extraction process decomposes oil shale and converts its kerogen into shale oil—a petroleum-like synthetic crude oil. The process is conducted by pyrolysis, hydrogenation, or thermal dissolution. The efficiencies of extraction processes are often evaluated by comparing their yields to the results of a Fischer Assay performed on a sample of the shale.

The oldest and the most common extraction method involves pyrolysis (also known as *retorting* or destructive distillation). In this process, oil shale is heated in the absence of oxygen until its kerogen decomposes into condensable shale oil vapours and non-condensable combustible oil shale gas. Oil vapours and oil shale gas are then collected and cooled, causing the shale oil to condense.

In addition, oil shale processing produces spent oil shale, which is a solid residue. Spent shale consists of inorganic compounds (minerals) and char (some authors use the terms *coke residue* or *semi-coke* instead of *char*)—a carbonaceous residue formed from kerogen. Burning the char off the spent shale produces oil shale ash. Spent shale and shale ash can be used as ingredients in cement or brick manufacture. The composition of the oil shale may lend added value to the extraction process through the recovery of by-products, including ammonia, sulfur, aromatic compounds, pitch, asphalt, and waxes.

Heating the oil shale to pyrolysis temperature and completing the endothermic kerogen decomposition reactions require a source of energy. Some technologies burn other fossil fuels such as natural gas, oil, or coal to generate this heat and experimental methods have used electricity, radio waves, microwaves, or reactive fluids for this purpose. Two strategies are used to reduce, and even eliminate, external heat energy requirements: the oil shale gas and char by-products generated by pyrolysis may be burned as a source of energy, and the heat contained in hot spent oil shale and oil shale ash may be used to pre-heat the raw oil shale.

For *ex situ* processing, oil shale is crushed into smaller pieces, increasing surface area for better extraction. The temperature at which decomposition of oil shale occurs depends on the time-scale of the process. In *ex situ* retorting processes, it begins at 300 °C (570 °F) and proceeds more rapidly and completely at higher temperatures. The amount of oil produced is the highest when the temperature ranges between 480 and 520 °C (900 and 970 °F).

The ratio of oil shale gas to shale oil generally increases along with retorting temperatures. For a modern *in situ* process, which might take several months of heating, decomposition may be conducted at temperatures as low as 250 °C (480 °F). Temperatures below 600 °C (1,110 °F) are preferable, as this prevents the decomposition of lime stone and dolomite in the rock and thereby limits carbon dioxide emissions and energy consumption.

Hydrogenation and thermal dissolution (reactive fluid processes) extract the oil using hydrogen donors, solvents, or a combination of these. Thermal dissolution involves the application of solvents at elevated temperatures and pressures, increasing oil output by cracking the dissolved organic matter. Different methods produce shale oil with different properties.

Classification of Extraction Technologies

Industry analysts have created several classifications of the technologies used to extract shale oil from oil shale.

***By process principles*:** Based on the treatment of raw oil shale by heat and solvents the methods are classified as pyrolysis, hydrogenation, or thermal dissolution.

***By location*:** A frequently used distinction considers whether processing is done above or below ground, and classifies the technologies broadly as *ex situ* (displaced) or *in situ* (in place). In *ex situ* processing, also known as above-ground retorting, the oil shale is mined either underground or at the surface and then transported to a processing facility. In contrast, *in situ* processing converts the kerogen while it is still in the form of an oil shale deposit, following which it is then extracted via oil wells, where it rises in the same way as conventional crude oil. Unlike *ex situ* processing, it does not involve mining or spent oil shale disposal aboveground as spent oil shale stays underground.

***By heating method*:** The method of transfering heat from combustion products to the oil shale may be classified as direct or indirect. While methods that allow combustion products to contact the oil shale within the retort are classified as *direct*, methods that burn materials external to the retort to heat another material that contacts the oil shale are described as *indirect*

***By heat carrier*:** Based on the material used to deliver heat energy to the oil shale, processing technologies have been classified into gas heat carrier, solid heat carrier, wall conduction, reactive fluid, and volumetric heating methods. Heat carrier methods can be sub-classified as direct or indirect.

***By raw oil shale particle size*:** The various *ex situ* processing technologies may be differentiated by the size of the oil shale particles that are fed into the retorts. As a rule, gas heat carrier technologies process oil shale lumps varying in diameter from 10 to 100 millimetres (0.4 to 3.9 in), while solid heat carrier and wall conduction technologies process fines which are particles less than 10 millimetres (0.4 in) in diameter.

***By retort orientation*:** "Ex-situ" technologies are sometimes classified as vertical or horizontal. Vertical retorts are usually shaft kilns where a bed of shale moves from top to bottom by gravity. Horizontal retorts are usually horizontal rotating drums or screws where shale moves from one end to the other. As a general rule, vertical retorts process lumps using a gas heat carrier, while horizontal retorts process fines using solid heat carrier.

***By complexity of technology*:** *In situ* technologies are usually classified either as *true in situ* processes or *modified in situ* processes. *True in situ* processes do not involve mining or crushing the oil shale. *Modified in situ* processes involve drilling and fracturing the target oil shale deposit to create voids in the deposit. The voids enable a better flow of gases and fluids through the deposit, thereby increasing the volume and quality of the shale oil produced.

Chapter 3

Exploration and Production

Blowout (Well Drilling)

A blowout is the uncontrolled release of crude oil and/or natural gas from an oil well or gas well after pressure control systems have failed.

Prior to the advent of pressure control equipment in the 1920s, the uncontrolled release of oil and gas from a well while drilling was common and was known as an oil gusher, gusher or wild well.

History

Gushers were an icon of oil exploration during the late 19th and early 20th centuries. During that era, the simple drilling techniques such as cable-tool drilling and the lack of blowout preventers meant that drillers could not control high-pressure reservoirs.

When these high pressure zones were breached the hydrocarbon fluids would travel up the well at a high rate, forcing out the drill string and creating a gusher. A well which began as a gusher was said to have "blown in": for instance, the Lakeview Gusher *blew in* in 1910. These uncapped wells could produce large amounts of oil, often shooting 200 feet (60 m) or higher into the air. A blowout primarily composed of natural gas was known as a *gas gusher*.

Despite being symbols of new-found wealth, gushers were dangerous and wasteful. They killed workmen involved in drilling, destroyed equipment, and coated the landscape with thousands of barrels of oil; additionally, the explosive concussion released by the well when it pierces an oil/gas reservoir has been responsible for a number of oilmen losing their hearing entirely; standing too near to the drill at the moment it contacts the oil reservoir is extremely hazardous. The impact on wildlife is very hard to quantify, but can only be estimated to be mild in the most optimistic models — realistically, the ecological impact is estimated by scientists across the ideological spectrum to be severe, profound, and lasting.

To complicate matters further, the free flowing oil was — and is— in danger of igniting. One dramatic account of a blowout and fire reads,

> *"With a roar like a hundred express trains racing across the countryside, the well blew out, spewing oil in all directions. The derrick simply evaporated. Casings wilted like lettuce out of water, as heavy machinery writhed and twisted into grotesque shapes in the blazing inferno."*

The development of rotary drilling techniques where the density of the drilling fluid is sufficient to overcome the downhole pressure of a newly penetrated zone meant that gushers became avoidable. If however the fluid density was not adequate or fluids were lost to the formation, then there was still a significant risk of a well blowout.

In 1924 the first successful blowout preventer was brought to market. The BOP valve affixed to the wellhead could be closed in the event of drilling into a high pressure zone, and the well fluids contained. Well control techniques could be used to regain control of the well. As the technology developed, blowout preventers became standard equipment, and gushers became a thing of the past.

In the modern petroleum industry, uncontrollable wells became known as blowouts and are comparatively rare. There has been a significant improvement in technology, well control techniques and personnel training that has helped to prevent them occurring. From 1976 to 1981, 21 blowout reports are available.

Notable Gushers

1. The Shaw Gusher in Oil Springs, Ontario, was Canada's (and possibly the world's) first oil gusher. On January 16, 1862, it shot oil from over 60 metres (200 ft) below ground to above the treetops at a rate of 3,000 barrels (480 m^3) per day, triggering the oil boom in Lambton County.
2. Lucas Gusher at Spindletop in Beaumont, Texas in 1901 flowed at 100,000 barrels (16,000 m^3) per day at its peak, but soon slowed and was capped within nine days. The well tripled U.S. oil production overnight and marked the start of the Texas oil industry.

3. Masjed Soleiman, Iran in 1908 marked the first major oil strike recorded in the Middle East.
4. Lakeview Gusher on the Midway-Sunset Oil Field in Kern County, California of 1910 is believed to be the largest-ever U.S. gusher. At its peak, more than 100,000 barrels (16,000 m^3) of oil per day flowed out, reaching as high as 200 feet (60 m) in the air. It remained uncapped for 18 months, spilling over 9 million barrels (1,400,000 m^3) of oil, less than half of which was recovered.
5. A short-lived gusher at Alamitos #1 in Signal Hill, California in 1921 marked the discovery of the Long Beach Oil Field, one of the most productive oil fields in the world.
6. The Barroso 2 well in Cabimas, Venezuela in December 1922 flowed at around 100,000 barrels (16,000 m^3) per day for nine days, plus a large amount of natural gas.
7. Baba Gurgur near Kirkuk, Iraq, an oilfield known since antiquity, erupted at a rate of 95,000 barrels (15,100 m^3) a day in 1927.
8. The Wild Mary Sudik gusher in Oklahoma City, Oklahoma in 1930 flowed at a rate of 72,000 barrels (11,400 m^3) per day.
9. The Daisy Bradford gusher in 1930 marked the discovery of the East Texas Oil Field, the largest oilfield in the contiguous United States.
10. The largest known 'wildcat' oil gusher blew near Qom, Iran on August 26, 1956. The uncontrolled oil gushed to a height of 52 m (170 ft), at a rate of 120,000 barrels (19,000 m^3) per day. The gusher was closed after 90 days' work by Bagher Mostofi and Myron Kinley (USA).
11. One of the most troublesome gushers happened on June 23, 1985 at the well #37 at the Tengiz field in Atyrau, Kazakh SSR, Soviet Union, where the deep, 4209 metre well blew out and the 200-metres high gusher self-ignited two days later. Oil pressure up to 800 atm and high hydrogen sulfide content had led to the gusher being capped only on 27 July, 1986 when the well was closed by the shaped charge. The total volume of erupted material measured at 4.3 millions metric tons of oil, 1.7 bn m^3 of natural gas, and the burning gusher resulted in 890 tons of various mercaptans and more than 900000 tons of soot released into atmosphere.
12. The largest underwater blowout in U.S. history occurred on April 20, 2010, in the Gulf of Mexico at the Macondo Prospect oil field. The blowout caused the explosion of the Deepwater

Horizon, a mobile offshore drilling platform owned by Transocean and under lease to BP at the time of the blowout. While the exact volume of oil spilled is unknown, as of June 3, 2010 (2010 -06-03), the United States Geological Survey (USGS) Flow Rate Technical Group has placed the estimate at between 35,000 to 60,000 barrels (5,600 to 9,500 m^3) of crude oil per day.

Cause of Blowouts

Reservoir Pressure

Petroleum or crude oil is a naturally occurring, flammable liquid consisting of a complex mixture of hydrocarbons of various molecular weights, and other organic compounds, that are found in geologic formations beneath the Earth's surface.

Because most hydrocarbons are lighter than rock or water, they often migrate upward through adjacent rock layers until either reaching the surface or becoming trapped within porous rocks (known as reservoirs) by impermeable rocks above.

However, the process is influenced by underground water flows, causing oil to migrate hundreds of kilometres horizontally or even short distances downward before becoming trapped in a reservoir. When hydrocarbons are concentrated in a trap, an oil field forms, from which the liquid can be extracted by drilling and pumping. The down hole pressures experienced at the rock structures change depending upon the depth and the characteristic of the source rock.

The downhole fluid pressures are controlled in modern wells through the balancing of the hydrostatic pressure provided by the mud used. Should the balance of the drilling mud pressure be incorrect then formation fluids (oil, natural gas and/or water) begin to flow into the wellbore and up the annulus (the space between the outside of the drill string and the walls of the open hole or the inside of the last casing string set), and/or inside the drill pipe.

This is commonly called a *kick*. If the well is not shut in (common term for the closing of the blow-out preventer valves), a kick can quickly escalate into a blowout when the formation fluids reach the surface, especially when the influx contains gas that expands rapidly as it flows up the wellbore, further decreasing the effective weight of the fluid.

Additional mechanical barriers such as blowout preventers (BOPs) can be closed to isolate the well while the hydrostatic balance is regained through circulation of fluids in the well.

Early warning signs of a well kick are:

- Sudden change in drilling rate;
- Change in surface fluid rate;
- Change in pump pressure;
- Reduction in drillpipe weight;
- Surface mud cut by gas, oil or water.

The primary means of detecting a kick is a relative change in the circulation rate back up to the surface into the mud pits. The drilling crew or mud engineer keeps track of the level in the mud pits and/or closely monitors the rate of mud returns versus the rate that is being pumped down the drill pipe.

Upon encountering a zone of higher pressure than is being exerted by the hydrostatic head of the drilling mud at the bit, an increase in mud returns would be noticed as the formation fluid influx pushes the drilling mud toward the surface at a higher rate.

Conversely, if the rate of returns is slower than expected, it means that a certain amount of the mud is being lost to a thief zone somewhere below the last casing shoe. This does not necessarily result in a kick (and may never become one); however, a drop in the mud level might allow influx of formation fluids from other zones if the hydrostatic head at depth is reduced to less than that of a full column of mud.

Well Control

The first response to detecting a kick would be to isolate the wellbore from the surface by activating the blow-out preventers and closing in the well. Then the drilling crew would attempt to circulate in a heavier *kill fluid* to increase the hydrostatic pressure (sometimes with the assistance of a well control company).

In the process, the influx fluids will be slowly circulated out in a controlled manner, taking care not to allow any gas to accelerate up the wellbore too quickly by controlling casing pressure with chokes on a predetermined schedule.

This effect will be minor if the influx fluid is mainly salt water. And with an oil-based drilling fluid it can be masked in the early stages of controlling a kick because gas influx may dissolve into the oil under pressure at depth, only to come out of solution and expand rather rapidly as the influx nears the surface.

Once all the contaminant has been circulated out, the casing pressure should have reached zero.

Types of Blowouts

Surface Blowouts

Blowouts can eject the drill string out of the well, and the force of the escaping fluid can be strong enough to damage the drilling rig.

In addition to oil, the output of a well blowout might include sand, mud, rocks, drilling fluid, natural gas, water, and other substances. Blowouts will often be ignited by an ignition source, from sparks from rocks being ejected, or simply from heat generated by friction.

A well control company will then need to extinguish the well fire or cap the well, and replace the casing head and hangars. The flowing gas may contain poisonous hydrogen sulfide and the oil operator might decide to ignite the stream to convert this to less hazardous substances.

Sometimes, blowouts can be so forceful that they cannot be directly brought under control from the surface, particularly if there is so much energy in the flowing zone that it does not deplete significantly over the course of a blowout.

In such cases, other wells (called relief wells) may be drilled to intersect the well or pocket, in order to allow kill-weight fluids to be introduced at depth. When first drilled in the 1930s relief well were drilled to inject water into the main drill well hole.

Contrary to what might be inferred from the term, such wells generally are not used to help relieve pressure using multiple outlets from the blowout zone.

Subsea Blowouts

Subsea wells have the wellhead and pressure control equipment located on the seabed. They vary from depths of 10 feet (3.0 m) to 8,000 feet (2,400 m). It is very difficult to deal with a blowout in very deep water because of the remoteness and limited experience with this type of situation.

The Deepwater Horizon well blowout in the Gulf of Mexico in April 2010, in 5,000 feet (1,500 m) water depth, is the deepest subsea well blowout to date.

Underground Blowouts

An underground blowout is a special situation where fluids from high pressure zones flow uncontrolled to lower pressure zones within the wellbore. Usually this is from deeper higher pressure zones to shallower lower pressure formations.

There may be no escaping fluid flow at the wellhead. Underground blowouts can be very difficult to bring under control, and if left unchecked the fluids may find their way to the surface or ocean floor nearby.

Blowout Control Expertise

Myron M. Kinley was a pioneer in fighting oil well fires and blowouts. He developed many patents and designs for the tools and techniques of oil firefighting. His father, Karl T. Kinley, attempted to extinguish an oil well fire with the help of a massive explosion — a method that remains a common technique for fighting oil fires.

The first oil well put out with explosives by Myron Kinley and his father, was in 1913. Kinley would later form the M.M. Kinley Company in 1923.

Paul N. "Red" Adair joined the M.M. Kinley Company in 1946, and worked 14 years with Myron Kinley before starting his own company, Red Adair Co., Inc., in 1959.

Asger "Boots" Hansen and Edward Owen "Coots" Matthews also begin their careers under Kinley.

Red Adair co. has helped in controlling many offshore blowouts, including;

- CATCO fire in the Gulf of Mexico in 1959
- "The Devil's Cigarette Lighter" in 1962 in Gassi Touil, Algeria, in the Sahara Desert
- The Ixtoc I oil spill in Mexico's Bay of Campeche in 1979
- The Piper Alpha disaster in the North Sea in 1988
- The Kuwaiti oil fires following the Gulf War in 1991.

In 1994, Adair retired and sold his company to Global Industries. Management of Adair's company left and created International Well Control (IWC). In 1997, they would buy the company Boots & Coots International Well Control, Inc., which was founded by two former lieutenants of Red Adair in 1978.

Methods of Quenching Blowouts

Although several experimental methods exist which attempt to capture as much oil as possible from a blown out well, they are very far from perfect, capturing between 20% - 50% of the leaking oil, by optimistic estimates.

Ideally, the well could be made to stop gushing oil entirely - thus putting a stop to the cumulating pollution.

Notable Offshore Well Blowouts

Table : *Data from industry information*

Year	*Rig Name*	*Rig Owner*	*Type*	*Damage/details*
1955	S-44	Chevron Corporation	Sub Recessed pontoons	Blowout and fire. Returned to service.
1959	C. T. Thornton	Reading & Bates	Jackup	Blowout and fire damage.
1964	C. P. Baker	Reading & Bates	Drill barge	Blowout in Gulf of Mexico, vessel capsized, 22 killed.
1965	Trion	Royal Dutch Shell	Jackup	Destroyed by blowout.
1965	Pagur	SNAM	Jackup	Destroyed by blowout and fire.
1968	Little Bob	Coral	Jackup	Blowout and fire, killed 7.
1969	Wodeco III	Floor drilling	Drilling barge	Blowout
1969	Sedco 135G	Sedco Inc	Semi-submersible	Blowout damage
1969	Rimrick Tidelands	ODEC	Submersible	Blowout in Gulf of Mexico
1970	Stormdrill III	Storm Drilling	Jackup	Blowout and fire damage.
1970	Discoverer III	Offshore Co.	Drillship	Blowout (S. China Seas)
1971	Big John	Atwood Oceanics	Drill barge	Blowout and fire.
1971	Unknown	Floor Drilling	Drill barge	Blowout and fire off Peru, 7 killed.
1972	J. Storm II	Marine Drilling Co.	Jackup	Blowout in Gulf of Mexico
1972	M. G. Hulme	Reading & Bates	Jackup	Blowout and capsize in Java Sea.
1972	Rig 20	Transworld Drilling	Jackup	Blowout in Gulf of Martaban.
1973	Mariner I	Sante Fe Drilling	Semi-sub	Blowout off Trinidad, 3 killed.
1975	Mariner II	Sante Fe Drilling	Semi-submersible	Lost BOP during blowout.
1975	J. Storm II	Marine Drilling Co.	Jackup	Blowout in Gulf of Mexico.
1976	Petrobras III	Petrobras	Jackup	No info.
1976	W. D. Kent	Reading & Bates	Jackup	Damage while drilling relief well.
1977	Maersk Explorer	Maersk Drilling	Jackup	Blowout and fire in North Sea
1977	Ekofisk Brav	Phillips Petroleum	Platform	Blowout during well workover.
1978	Scan Bay	Scan Drilling	Jackup	Blowout and fire in the Persion Gulf.
1979	Salenergy II	Salen Offshore	Jackup	Blowout in Gulf of Mexico

Contd...

Year	***Rig Name***	***Rig Owner***	***Type***	***Damage/details***
1979	Sedco 135F	Sedco Drilling	Semi-submersible	Blowout and fire in Bay of Campeche Ixtoc I well.
1980	Sedco 135G	Sedco Drilling	Semi-submersible	Blowout and fire of Nigeria.
1980	Discoverer 534	Offshore Co.	Drillship	Gas escape caught fire.
1980	Ron Tappmeyer	Reading & Bates	Jackup	Blowout in Persian Gulf, 5 killed.
1980	Nanhai II	Peoples Republic of China	Jackup	Blowout of Hainan Island
1980	Maersk Endurer	Maersk Drilling	Jackup	Blowout in Red Sea, 2 killed.
1980	Ocean King	ODEC	Jackup	Blowout and fire in Gulf of Mexico, 5 killed.
1980	Marlin 14	Marlin Drilling	Jackup	Blowout in Gulf of Mexico
1981	Penrod 50	Penrod Drilling	Submersible	Blowout and fire in Gulf of Mexico.
1985	West Vanguard	Smedvig	Semi-submersible	Shallow gas blowout and fire in Norwegian sea, 1 fatality.
1981	Petromar V	Petromar	Drillship	Gas blowout and capsize in S. China seas.
1988	Ocean Odyssey	Diamond Offshore Drilling	Semi-submersible	Gas blowout at BOP and fire in the UK North Sea, 1 killed.
1989	Al Baz	Sante Fe	Jackup	Shallow gas blowout and fire in Nigeria, 5 killed.
1993	Actinia	Transocean	Semi-submersible	Sub-sea blowout in Vietnam..
2001	Ensco 51	Ensc	Jackup	Gas blowout and fire, Gulf of Mexico, no casualties
2002	Arabdrill 19	Arabian Drilling Co.	Jackup	Structural collapse, blowout, fire and sinking.
2004	Adriatic IV	Global Sante Fe	Jackup	Blowout and fire at Temsah platform, Mediterranean Sea
2007	Usumacinta	PEMEX	Jackup	Storm force rig to move, causing well blowout on Kab 101 platform, 22 killed.
2009	West Atlas/Montara	Seadrill	Jackup/Platform	Blowout and fire on rig and platform in Australia.
2010	Deepwater Horizon	Transocean	Semi-submersible	Blowout and fire on the rig, subsea well blowout, killed 11 in explosion.
2010	Vermilion Block 380	Mariner Energy	Platform	Blowout and fire, 13 survivors, 1 injured.

Use of Nuclear Explosions

On Sep. 30, 1966 the Soviet Union in Urt-Bulak, an area about 80 kilometres from Bukhara, Uzbekistan, experienced blowouts on five natural gas wells. It was claimed in Komsomoloskaya Pravda that after years of burning uncontrollably they were able to stop them entirely.

The Soviets lowered a specially made 30 kiloton nuclear bomb into a 6 kilometres (20,000 ft) borehole drilled 25 to 50 metres (82 to 160 ft) away from the original (rapidly leaking) well. A nuclear explosive was deemed necessary because conventional explosive both lacked the necessary power and would also require a great deal more space underground. When the bomb was set off, it proceeded to crush the original pipe that was carrying the gas from the deep reservoir to the surface, as well as to glassify all the surrounding rock. This caused the leak and fire at the surface to cease within approximately one minute of the explosion, and proved over the years to have been a permanent solution. A second attempt on a similar well was not as successful and other tests were for such experiments as oil extraction enhancement (Stavropol, 1969) and the creation of gas storage reservoirs (Orenburg, 1970). As nuclear historian Robert S. Norris notes in the Times, all these Soviet nuclear blasts were on land and did not involve oil. That being the case, the general principles involved are the same.

Completion (Oil and Gas Wells)

In petroleum production, completion is the process of making a well ready for production (or injection). This principally involves preparing the bottom of the hole to the required specifications, running in the production tubing and its associated down hole tools as well as perforating and stimulating as required. Sometimes, the process of running in and cementing the casing is also included.

Lower Completion

This refers to the portion of the well across the production or injection zone. The well designer has many tools and options available to design the lower completion according to the conditions of the reservoir. Typically, the lower completion is set across the productive zone using a liner hanger system, which anchors the lower completion to the production casing string. The broad categories of lower completion are listed below.

Barefoot Completion

Selective isolation of oil, gas and barefoot section. Its an open hole, but with simple design and low cost. can be used in consolidated formations. It's the technique used by drilling companies in deep wells, Natural fractured reservoirs and Multilateral.

Open Hole Completion

This designation refers to a range of completions where no casing or liner is cemented in place across the production zone. In competent formations the zone might be left entirely bare, but usually some sort of sand-control and/or flow-control means are incorporated.

Openhole completions have seen significant uptake in recent years, and there are many configurations, often developed to address specific reservoir challenges. There have been many recent developments that have boosted the success of openhole completions, and they also tend to be popular in horizontal wells, where cemented installations are more expensive and technically more difficult. The common options for openhole completions are; Pre-holed liner (also often called pre-drilled liner). The liner is prepared with multiple small drilled holes, then set across the production zone to provide wellbore stability and an intervention conduit.

Pre-holed liner is often combined with openhole packers, such as swelling elastomers, mechanical packers or external casing packers, to provide zonal segregation and isolation. It is now quite common to see a combination of pre-holed liner, solid liner and swelling elastomer packers to provide an initial isolation of unwanted water or gas zones. Multiple sliding sleeves can also be used in conjunction with openhole packers to provide considerable flexibility in zonal flow control for the life of the wellbore. This type of completion is also being adopted in some water injection wells, although these require a much greater performance envelope for openhole packers, due to the considerable pressure and temperature changes that occur in water injectors.

Openhole completions (in comparison with cemented pipe) require better understanding of formation damage, wellbore clean-up and fluid loss control. A key difference is that perforating penetrates through the first 6-18 inches (15-45 cm) of formation around the wellbore, whilst openhole completions require the reservoir fluids to flow through all of the filtrate-invaded zone around the wellbore and lift-off of the mud filter cake.

Many openhole completions will incorporate fluid loss valves at the top of the liner to provide well control whilst the upper completion is run. There are an increasing number of ideas coming into the market place to extend the options for openhole completions; for example, electronics can be used to actuate a self-opening or self-closing liner valve. This might be used in an openhole completion to improve clean-up, by bringing the well onto production from the toe-end for 100 days, then self-opening the heel-end. Inflow control devices and intelligent completions are also installed as openhole completions.

Pre-holed liner may provide some basic control of solids production, where the wellbore is thought to fail in aggregated chunks of rubble, but it is not typically regarded as a sand control completion.

Slotted liner can be selected as an alternative to pre-holed liner, sometimes as a personal preference or from established practice on a field. It can also be selected to provide a low cost control of sand/solids production. The slotted liner is machined with multiple longitudinal slots, for example 2 mm x 50mm, spread across the length and circumference of each joint. Recent advances in laser cutting means that slotting can now be done much cheaper to much smaller slot widths and in some situation slotted liner is now used for the same functionality as sand control screens.

Openhole sand control is selected where the liner is required to mechanically hold-back the movement of formation sand. There are many variants of openhole sand control, the three popular choices being stand-alone screens, openhole gravel packs (also known as external gravel packs, where a sized sand 'gravel' is placed as an annulus around the sand control screen) and expandable screens. Screen designs are mainly wire-wrap or premium; wire-wrap screens use spiral-welded corrosion-resistant wire wrapped around a drilled basepipe to provide a consistent small helical gap (such as 0.012-inch (0.30 mm), termed 12 gauge). Premium screens use a woven metal cloth wrapped around a basepipe. Expandable screens are run to depth before being mechanically swaged to a larger diameter. Ideally, expandable screens will be swaged until they contact the wellbore wall.

Cased Hole Completion

This involves running casing or a liner down through the production zone, and cementing it in place. Connection between the well bore and the formation is made by perforating. Because perforation intervals can be precisely positioned, this type of completion affords good control of fluid flow, although it relies on the quality of the cement to prevent fluid flow behind the liner. As such it is the most common form of completion...

Completion Components

The upper completion refers to all components from the bottom of the production tubing upwards. Proper design of this "completion string" is essential to ensure the well can flow properly given the reservoir conditions and to permit any operations as are deemed necessary for enhancing production and safety.

Wellhead

This is the pressure containing equipment at the surface of the well where casing strings are suspended and the Blowout preventer or Christmas tree (oil well) is connected.

Christmas Tree

This is the main assembly of valves that controls flow from the well to the process plant (or the other way round for injection wells) and allows access for chemical squeezes and well interventions.

Tubing Hanger

This is the component, which sits on top of the wellhead and serves as the main support for the production tubing.

Production Tubing

Production tubing is the main conduit for transporting hydrocarbons from the reservoir to surface (or injection material the other way). It runs from the tubing hanger at the top of the wellhead down to a point generally just above the top of the production zone.

Downhole Safety Valve

This component is intended as a last resort method of protecting the surface from the uncontrolled release of hydrocarbons. It is a cylindrical valve with either a ball or flapper closing mechanism. It is installed in the production tubing and is held in the open position by a high-pressure hydraulic line from surface contained in a 6.35 mm (1/4-) control line that is attached to the DHSV's hydraulic chamber and terminated at surface to an hydraulic actuator. The high pressure is needed to overcome the production pressure in the tubing upstream of the choke on the tree. The valve will operate if the umbilical HP line is cut or the wellhead/tree is destroyed.

This valve allows fluids to pass up or be pumped down the production tubing. When closed the DHSV forms a barrier in the direction of hydrocarbon flow, but fluids can still be pumped down for well kill operations. It is placed as far below the surface as is deemed safe from any possible surface disturbance including cratering caused by the wipeout of the platform.

Where hydrates are likely to form (most production is at risk of this), the depth of the SCSSV (surface-controlled sub-surface safety valve) below the seabed may be as much as 1 km: this will allow for the geothermal temperature to be high enough to prevent hydrates from blocking the valve.

Annular Safety Valve

On wells with gas lift capability, many operators consider it prudent to install a valve, which will isolate the 'A' annulus for the same reasons a DHSV may be needed to isolate the production tubing in order to prevent the inventory of natural gas downhole from becoming a hazard as it became on Piper Alpha.

Side Pocket Mandrel

This is a welded/machined product which contains a 'side-pocket' alongside the main tubular conduit. The side pocket, typically 1⁻ or 1½⁻ diameter is designed to contain gas lift valve, which allows hydrocarbon gas from the 'A' annulus to be injected into the flow stream.

Electrical Submersible Pump

This device is used for artificial lift to help provide energy to drive hydrocarbons to surface if reservoir pressure is insufficient.

- Artificial Lift - Baker Hughes Last accessed 20-Jun-2011.

Landing Nipple

This is a receptacle to receive wireline tools. It is also a useful marker for depths in the well, which can be difficult to accurately determine. Normally it is set close to the end of the tubing string to be able to isolate the same from the reservoir conditions, at any time during the producing life of the well.

Sliding Sleeve

The sliding sleeve is hydraulically or mechanically actuated to allow communication between the tubing and the 'A' annulus. They are often used in multiple reservoir wells to regulate flow to and from the zones.

Production Packer

The packer isolates the annulus between the tubing and the inner casing and the foot of the well. This is to stop reservoir fluids from flowing up the full length of the casing and damaging it. It is generally placed close to the foot of the tubing, shortly above the production zone.

Downhole Gauges

This is an electronic or fibre optic sensor to provide continuous monitoring of downhole pressure and temperature. Gauges use a 1/4⁻

control line clamped onto the outside of the tubing string to provide an electrical or fibre optic communication to surface.

Perforated Joint

This is a length of tubing with holes punched into it. If used, it will normally be positioned below the packer and will offer an alternative entry path for reservoir fluids into the tubing in case the shoe becomes blocked, for example, by a stuck perforation gun.

Formation Isolation Valve

This component, placed towards the foot of the completion string, is used to provide two way isolation from the formation for completion operations without the need for kill weight fluids. Their use is sporadic as they do not enjoy the best reputation for reliability when it comes to opening them at the end of the completion process.

Centralizer

In highly deviated wells, this components may be included towards the foot of the completion. It consists of a large collar, which keeps the completion string centralised within the hole.

Wireline Entry Guide

This component is often installed at the end of the tubing (the shoe). It is intended to make pulling out wireline tools easier by offering a guiding surface for the toolstring to re-enter the tubing without getting caught on the side of the shoe.

- Schlumberger WEG (pdf) Last accessed: 15-Jun-2011.

Perforating and Stimulating

In cased hole completions (the majority of wells), once the completion string is in place, the final stage is to make a connection between the wellbore and the formation. This is done by running perforation guns to blast holes in the casing or liner to make a connection. Modern perforations are made using shaped explosive charges, similar to the armor-penetrating charge used on antitank rockets (bazookas). Sometimes once the well is fully completed, further stimulation is necessary to achieve the planned productivity. There are a number of stimulation techniques.

Acidizing

This involves the injection of chemicals to eat away at any skin damage, "cleaning up" the formation, thereby improving the flow of reservoir fluids. Acid can also be used to clean the wellbore of some scales that form from mineral laden produced water.

Fracturing

This means creating and extending fractures from the perforation tunnels deeper into the formation, increasing the surface area for formation fluids to flow into the well, as well as extending past any possible damage near the wellbore. This may be done by injecting fluids at high pressure (hydraulic fracturing), injecting fluids laced with round granular material (proppant fracturing), or using explosives to generate a high pressure and high speed gas flow (TNT or PETN up to 1,900,000 psi (13,000,000 kPa)) and (propellant stimulation up to 4,000 psi (28,000 kPa)).

Acidising and Fracturing (Combined Method)

This involves use of explosives and injection of chemicals to increase acid-rock contact.

Nitrogen Circulation

Sometimes, productivity may be hampered due to the residue of completion fluids, heavy brines, in the wellbore. This is particularly a problem in gas wells. In these cases, coiled tubing may be used to pump nitrogen at high pressure into the bottom of the borehole to circulate out the brine.

Wellhead

A wellhead is a general term used to describe the component at the surface of an oil or gas well that provides the structural and pressure-containing interface for the drilling and production equipment.

The primary purpose of a wellhead is to provide the suspension point and pressure seals for the casing strings that run from the bottom of the hole sections to the surface pressure control equipment.

While drilling the oil well, surface pressure control is provided by a blowout preventer (BOP). If the pressure is not contained during drilling operations by the column of drilling fluid, casings, wellhead, and BOP, a well blowout could occur. Once the well has been drilled, it is *completed* to provide an interface with the reservoir rock and a tubular conduit for the well fluids. The surface pressure control is provided by a christmas tree, which is installed on top of the wellhead, with isolation valves and choke equipment to control the flow of well fluids during production.

Wellheads are typically welded onto the first string of casing, which has been cemented in place during drilling operations, to form an integral structure of the well. In exploration wells that are later abandoned, the wellhead may be recovered for refurbishment and re-use. Offshore, where a wellhead is located on the production platform it is called a *surface wellhead*, and if located beneath the water then it is referred to as a subsea wellhead or mudline wellhead.

Functions

A wellhead serves numerous functions, some of which are:

1. Provide a means of casing suspension. (Casing is the permanently installed pipe used to line the well hole for pressure containment and collapse prevention during the drilling phase).
2. Provides a means of tubing suspension. (Tubing is removable pipe installed in the well through which well fluids pass).
3. Provides a means of pressure sealing and isolation between casing at surface when many casing strings are used.
4. Provides pressure monitoring and pumping access to annuli between the different casing/tubing strings.
5. Provides a means of attaching a blowout preventer during drilling.
6. Provides a means of attaching a christmas tree for production operations.
7. Provides a reliable means of well access.
8. Provides a means of attaching a well pump.

Components

The primary components of a wellhead system are:

- casing head
- casing spools
- casing hangers
- packoffs (isolation) seals
- bowl protectors/wear bushings
- test plugs
- mudline suspension systems
- tubing heads
- tubing hangers
- tubing head adapters.

Design Specification

The oil industry specifications for wellhead systems (materials, dimensions, test procedures and pressure ratings etc.) are :

1. API 6A Specification for Wellhead and Christmas Tree Equipment
2. ISO 10423 Wellhead and Christmas Tree Equipment.

Christmas Tree (Oil Well)

In petroleum and natural gas extraction, a Christmas tree, or "tree", (not "wellhead" as sometimes incorrectly referred to) is an assembly of valves, spools, and fittings used for an oil well, gas well, water injection well, water disposal well, gas injection well, condensate well and other types of wells. It was named for its crude resemblance to a decorated tree.

Note that a tree and wellhead are separate pieces of equipment not to be mistaken as the same piece. A wellhead must be present in order to utilize a Christmas tree and a wellhead is used *without* a Christmas tree during drilling operations, and also for riser tie-back situations which would then have a tree included at riser top. Producing surface wells that require pumps (pump jacks, nodding donkeys, and so on) frequently do not utilize any tree due to no pressure containment requirement.

Tree complexity has increased over the last few decades. They are frequently manufactured from blocks of steel containing multiple valves rather than made from multiple flanged valves. This is especially true in subsea applications where the resemblance to Christmas trees no longer exists given the frame and support systems into which the main valve block is integrated.

It is common to identify the type of tree as either "subsea tree" or "surface tree". Each of these classifications has a number or varieties within them. Examples of subsea include conventional, dual bore, mono bore, TFL (through flow line), horizontal, mudline, mudline horizontal, side valve, and TBT (through bore tree) trees.

The primary function of a tree is to control the flow into or out of the well, usually oil or gas. A tree often provides numerous additional functions including chemical injection points, well intervention means, pressure relief means (such as annulus vent), tree and well monitoring points (such as pressure, temperature, corrosion, erosion, sand detection, flow rate, flow composition, valve and choke position feedback, connection points for devices such as down hole pressure and temperature transducer (DHPT).

When the operator, well, and facilities are ready to produce and receive oil or gas, valves are opened and the release of the formation fluids is allowed to flow into and through a pipeline. The pipeline then leads to a processing facility, storage depot and/or other pipeline eventually leading to a refinery or distribution centre (for gas). Subsea wells and thus trees usually flow through flowlines to a fixed or floating production platform or to a storage vessel (known as a floating storage offloading vessel (FSO), or floating processing unit (FPU), or floating production and offloading vessel or FPSO or other combination of structures).

A tree may also be used to control the injection of gas or water injection application on a producing or non-producing well in order to sustain economic "production" volumes of oil from other well(s) in the area (field). On producing wells, injection of chemicals or alcohols or oil distillates to prevent and or solve production problems (such as blockages) may be used. Functionality may be extended further by using the control system on a subsea tree to monitor, measure, and react to sensor outputs on the tree or even down the well bore.

The control system attached to the tree controls the downhole safety valve (scssv, dhsv, sssv) while the tree acts as an attachment and conduit means of the control system to the downhole safety valve.

Christmas trees are used on both surface and subsea wells (current technical limits are up to around 3000 metres and working temperatures of -50°F to 350°F with a pressure of up to 15,000 psi). The deepest installed subsea tree is in the Gulf of Mexico at approximately 9,000 feet (2,700 m).

Valves

Subsea and surface trees have a large variety of valve configurations and combinations of manual and/or actuated (hydraulic or pneumatic) valves. Examples are identified in API Specifications 6A and 17D.

A basic surface tree consists of two or three manual valves (usually gate valves because of their strength).

A typical sophisticated surface tree will have at least four or five valves, normally arranged in a crucifix type pattern (hence the endurance of the term "Christmas tree"). The two lower valves are called the master valves (upper and lower respectively) because they lie in the flow path, which well fluids must take to get to surface. The lower master valve will normally be manually operated, while the upper master valve is often hydraulically actuated, allowing it to be a means of well control while an actuated wing valve is normally the primary well remotely (from control room or control panel) controlled valve.

Hydraulic tree wing valves are usually built to be fail safe closed, meaning they require active hydraulic pressure to stay open.

The right hand valve is often called the flow wing valve or the production wing valve, because it is in the flowpath the hydrocarbons take to production facilities (or the path water or gas will take from production to the well in the case of injection wells).

The left hand valve is often called the kill wing valve. It is primarily used for injection of fluids such as corrosion inhibitors or methanol to prevent hydrate formation. In the North Sea, it is called the non-active side arm (NASA). It is typically manually operated.

The valve at the top is called the swab valve and lies in the path used for well interventions like wireline and coiled tubing. For such operations, a lubricator is rigged up onto the top of the tree and the wire or coil is lowered through the lubricator, past the swab valve and into the well. This valve is typically manually operated.

Some trees have a second swab valve, the two arranged one on top of the other. The intention is to allow rigging down equipment from the top of the tree with the well flowing while still preserving the Two-barrier rule. With only a single swab valve, the upper master valve is usually closed to act as the second barrier, forcing the well to be shut in for a day during rig down operations. However, avoiding delaying production for a day is usually too small a gain to be worth the extra expense of a having a Christmas tree with a second swab valve.

Subsea trees are available in either vertical or horizontal configurations with further speciality available such as dual bore, monobore, concentric, drill-through, mudline, guidlineless or guideline. Subsea trees may range in size and weight from a few tons to approximately 70 tons for high pressure, deepwater (>3000 feet) guidelineless applications. Subsea trees contain many additional valves and accessories compared to Surface trees. Typically a subsea tree would have a choke (permits control of flow), a floline connection interface (hub, flange or other connection), subsea control interface (direct hydraulic, electro hydraulic, or electric) and sensors for gathering data such as pressure, temperature, sand flow, erosion, multiPhase flow, single phase flow such as water or gas.

Tubing Hanger

A tubing hanger is a component used in the completion of oil and gas production wells. It is set in the tree or the wellhead and suspends the production tubing and/or casing. Sometimes it provides porting to allow the communication of hydraulic, electric and other downhole functions, as well as chemical injection. It also serves to seal-in the annulus and production areas.

Production Tubing

Production tubing is a tube used in a wellbore through which production fluids are produced (travel).

Production tubing is run into the drilled well after the casing is run and cemented in place. Production tubing protects wellbore casing from wear, tear, corrosion, and deposition of by-products, such as sand/silt, paraffins, and asphaltenes. Along with other components that constitute the production string, it provides a continuous bore from the production zone to the wellhead through which oil and gas can be produced. It is usually between five and ten centimeters in diameter and is held inside the casing through the use of expandable packing devices. Purpose and design of production tubing is to enable quick, efficient, and safe installation, removal and re-installation.

If there is more than one zone of production in the well, up to four lines of production tubing can be run.

Downhole Safety Valve

A downhole safety valve refers to a component on an oil and gas well, which acts as a failsafe to prevent the uncontrolled release of reservoir fluids in the event of a worst case scenario surface disaster. It is almost always installed as a vital component on the completion.

Operation

These valves are commonly uni-directional flapper valves which open downwards, such that the flow of wellbore fluids try to push it shut, while pressure from surface pushes it open. This means that when closed, it will isolate the reservoir fluids from surface.

Most downhole safety valves are controlled hydraulically from surface, meaning they are opened using a hydraulic connection linked directly to a well control panel. When hydraulic pressure is applied down a control line, the hydraulic pressure forces a sleeve within the valve to slide downwards. This movement compresses a large spring and pushes the flapper downwards to open the valve. When hydraulic pressure is removed, the spring pushes the sleeve back up and causes the flapper to shut. In this way, it is failsafe and will isolate the wellbore in the event of a loss of the wellhead. The full designation for a typical valve is 'tubing retrievable, surface controlled, subsurface safety valve', abbreviated to TR-SCSSV.

Positioning

The location of the downhole safety valve within the completion is a precisely determined parameter intended to optimise safety. There are arguments against it either being too high or too low in the well and so the final depth is a compromise of all factors. MMS regulations state that the valve must be placed no less than 100' below the mudline.

Reasons to Keep it High

The further down the well the DHSV is located, the greater the potential inventory of hydrocarbons above it when closed. This means that in the event of loss of containment at surface, there is more fluid to be spilled causing environmental damage, or in the worst case, more fuel for a fire. Therefore, placing the valve higher limits this hazard.

Another reason relates to the hydraulic control line. Hydraulic pressure is required to keep the valve open as part of the failsafe design. However, if the valve is too far down the well, then the portion of the control line below the surface is larger. If it is too large, then the weight of the hydraulic fluid alone may apply sufficient pressure to keep the valve open, even with the loss of surface pressurisation.

Reasons to Keep it Low

As part of the role of the DHSV to isolate the surface from wellbore fluids, it is necessary for the valve to be positioned away from the well where it could potentially come to harm. This implies that it must be placed subsurface in all circumstances, i.e. in offshore wells, not above the seabed. There is also the risk of cratering in the event of a catastrophic loss of the topside facility. The valve is specifically placed below the maximum depth where cratering is expected to be a risk.

If there is a risk of methane hydrate (clathrate) plugs forming as the pressure changes through the valve due to Joule-Thomson cooling, then this is a reason to keep it low, where the rock is warmer than 30 C.

Deploying and Retrieving

Most downhole safety valves installed as part of the completion design are classed as "tubing retrievable". This means that they are installed as a component of the completion string and run in during completion. Retrieving the valve, should it malfunction, requires a workover. The full name for this most common type of downhole safety valve is a Tubing Retrievable Surface Controlled Sub-Surface Valve, shortened in completion diagrams to TRSCSSV.

If a tubing retrievable valve fails, rather than go to the expense of a workover, a "wireline retrievable" valve may be used instead. This type of valve can fit inside the production tubing and is deployed on wireline after the old valve has been straddled open.

Legal Requirement

The importance of DHSVs is undisputed. Graphic images of oil wells in Kuwait on fire after the First Gulf War after their wellheads were removed, demonstrate the perils of not using the components (at the time, they were deemed unnecessary because they were onshore wells). It is, however, not a direct legal requirement in many places. In the United Kingdom, no law mandates the use of DHSVs. However, the 1974 Health & Safety at Work Act requires that measures are taken to ensure that the uncontrolled release of wellbore fluids is prevented even in the worst case.

The brilliance of the act is that it does not issue prescriptive guideline for how to achieve the goal of health and safety, but merely sets out the requirement that the goal be achieved. It is up to the oil companies to decide how to achieve it and DHSVs are an important component of that decision. As such, although not a legal requirement, it is company policy for many operators in the UKCS.

Issues

While the DHSV isolates the production tubing, a loss of integrity could allow wellbore fluid to bypass the valve and escape to surface through the annulus. For wells using gas lift, it is often a requirement to install a safety valve in the 'A' annulus of the well to ensure that the surface is protected from a loss of annulus containment. However, these valves are not as common and they are not necessarily installed at the same position in the well, meaning it is possible that fluids could snake their way around the valves to surface.

Submersible Pump

A submersible pump (or electric submersible pump (ESP)) is a device which has a hermetically sealed motor close-coupled to the pump body. The whole assembly is submerged in the fluid to be pumped. The main advantage of this type of pump is that it prevents pump cavitation, a problem associated with a high elevation difference between pump and the fluid surface. Submersible pumps push fluid to the surface as opposed to jet pumps having to pull fluids. Submersibles are more efficient than jet pumps.

Working Principle

The submersible pumps used in ESP installations are multistage centrifugal pumps operating in a vertical position. Although their constructional and operational features underwent a continuous evolution over the years, their basic operational principle remained the same. Produced liquids, after being subjected to great centrifugal forces caused by the high rotational speed of the impeller, lose their kinetic energy in the diffuser where a conversion of kinetic to pressure energy takes place. This is the main operational mechanism of radial and mixed flow pumps.

The pump shaft is connected to the gas separator or the protector by a mechanical coupling at the bottom of the pump. Well fluids enter the pump through an intake screen and are lifted by the pump stages. Other parts include the radial bearings (bushings) distributed along the length of the shaft providing radial support to the pump shaft turning at high rotational speeds. An optional thrust bearing takes up part of the axial forces arising in the pump but most of those forces are absorbed by the protector's thrust bearing.

Applications

Submersible pumps are found in many applications. Single stage pumps are used for drainage, sewage pumping, general industrial pumping and slurry pumping. They are also popular with aquarium filters. Multiple stage submersible pumps are typically lowered down a borehole and used for water abstraction, water wells and in oil wells.

Special attention to the type of ESP is required when using certain types of liquids. ESP's commonly used on board naval vessels cannot be used to dewater contaminated flooded spaces. These use a 440 volt A/C motor that operates a small centrifugal pump. It can also be used out of the water, taking suction with a 2-1/2 inch non-collapsible hose. The pumped liquid is circulated around the motor for cooling purposes. There is a possibility that the gasoline will leak into the pump causing a fire or destroying the pump, so hot water and flammable liquids should be avoided.

ESP Usage in Oil Wells

Submersible pumps are used in oil production to provide a relatively efficient form of "artificial lift", able to operate across a broad range of flow rates and depths. By decreasing the pressure at the bottom of the well (by lowering bottomhole flowing pressure, or increasing drawdown), significantly more oil can be produced from the well when compared with natural production. The pumps are typically electrically powered and referred to as Electrical Submersible Pumps (ESP).

ESP systems consist of both surface components (housed in the production facility, for example an oil platform) and sub-surface components (found in the well hole). Surface components include the motor controller (often a variable speed controller), surface cables and transformers. Subsurface components typically include the pump, motor, seal and cables. A gas separator is sometimes installed.

The pump itself is a multi-stage unit with the number of stages being determined by the operating requirements. Each stage consists of a driven impeller and a diffuser which directs flow to the next stage of the pump. Pumps come in diameters from 90mm (3.5 inches) to 254mm (10 inches) and vary between 1 metre (3 ft) and 8.7 metres (29 ft) in length. The motor used to drive the pump is typically a three phase, squirrel cage induction motor, with a nameplate power rating in the range 7.5 kW to 560 kW (at 60 Hz).

New varieties of ESP can include a water/oil separator which permits the water to be reinjected into the reservoir without the need to lift it to the surface. There are at least 15 brands of oilfield esps used throughout the world. Until recently, ESPs had been highly costly to install due to the requirement of an electric cable downhole. This cable had to be wrapped around jointed tubing and connected at each joint. New coiled tubing umbilicals allow for both the piping and electric cable to deployed with a single conventional coiled tubing unit.

The ESP system consists of a number of components that turn a staged series of centrifugal pumps to increase the pressure of the well fluid and push it to the surface. The energy to turn the pump comes from a high-voltage (3 to 5 kV) alternating-current source to drive a special motor that can work at high temperatures of up to 300 °F (149 °C) and high pressures of up to 5,000 psi (34 MPa), from deep wells of up to 12,000 feet (3.7 km) deep with high energy requirements of up to about 1000 horsepower (750 kW). ESPs have dramatically lower efficiencies with significant fractions of gas, greater than about 10% volume at the pump intake. Given their high rotational speed of up to 4000 rpm (67 Hz) and tight clearances, they are not very tolerant of solids such as sand.

Sliding Sleeve

A sliding sleeve is a standard component for the completion of an oil or gas well. Their main uses are to shut off flow from one or more reservoir zones or to regulate pressure between zones. There are two main categories of sliding sleeves: open/close and choking. Open/close sleeves are shifted between a full open position and a closed position. They are used to shut off flow from a zone for economic reasons or to shut off a zone that is depleting or producing too much water. In multi-zone wells, they are used to regulate which zones to produce from and which ones to shut off. Mechanically actuated sleeves are simple and inexpensive but require actuation by a "lock," which must be run in the well on wireline or coiled tubing. Hydraulically actuated sleeves are more complicated but can be actuated from a small pump at surface. King sleeves can be used to regulate the pressure between two or more zones. They are also used to regulate the flow of fluid into a well during proppant fracturing or hydraulic fracturing operations. Choking sleeves are all hydraulically actuated and have a much more complex design than open/close sleeves.

A "sliding sleeve is also:

1. Part of a pencil for heavy-handed users.
2. Sliding sleeve supports the lead for precise lines and retracts as the pencil is used.
3. The sleeve is designed to protect the lead and minimize lead breakage.
4. Ideal for heavy-handed writers needing to use fine lead.

Production Packer

A production packer is a standard component of the completion hardware of oil or gas wells used to provide a seal between the outside of the production tubing and the inside of the casing, liner, or wellbore wall.

Based on its primary use, packers can be divided into two main categories: production packers and service packers. Production packers are those that remain in the well during well production. Service packers are used temporarily during well service activities such as cement squeezing, acidizing, fracturing and well testing.

It is usually run in close to the bottom end of the production tubing and set at a point above the top perforations or sand screens. In wells with multiple reservoir zones, packers are used to isolate the perforations for each zone. In these situations, a sliding sleeve would be used to select which zone to produce. Packers may also be used to protect the casing from pressure and produced fluids, isolate sections of corroded casing, casing leaks or squeezed perforations, and isolate or temporarily abandon producing zones. In waterflooding developments in which water is injected into the reservoir, packers are used in injection wells to isolate the zones into which the water must be injected. There are occasions in which running a packer may not be desirable. For example, high volume wells that are produced both up the tubing and annulus will not include a packer. Rod pumped wells are not normally run with packers because the associated gas is produced up the annulus. In general, well completions may not incorporate a packer when the annular space is used as a production conduit.

A production packer is designed to grip and seal against the casing ID. Gripping is accomplished with metal wedges called "slips." These components have sharpened, carburized teeth that dig into the metal of the casing. Sealing is accomplished with large, cylindrical rubber elements. In situations where the sealed pressure is very high (above 5,000 psi), metal rings are used on either side of the elements to prevent the rubber from extruding.

A packer is run in the casing on production tubing or wireline. Once the desired depth is reached, the slips and element must be expanded out to contact the casing. Axial loads are applied to push the slips up a ramp and to compress the element, causing it to expand outward. The axial loads are applied either hydraulically, mechanically, or with a slow burning chemical charge.

Most packers are "permanent" and require milling in order to remove them from the casing. The main advantages of permanent packers are lower cost and greater sealing and gripping capabilities.

In situations where a packer must be easily removed from the well, such as secondary recoveries, re-completions, or to change out the production tubing, a retrievable packer must be used. To unset the tool, either a metal ring is sheared or a sleeve is shifted to disengage connecting components. Retrievable packers have a more complicated design and generally lower sealing and gripping capabilities, but after removal and subsequent servicing, they can be reused.

Squeeze Job

Squeeze job, or squeeze cementing is a term often used in the oilfield to describe the process of injecting cement slurry into a zone, generally for pressure-isolation purposes. The term probably originated from the concept that enough water is "squeezed" out of the slurry to render it unflowable, so the portion that has actually entered the zone will stay in place when the squeeze pressure is removed. After surface indications (e.g., pressure reaching a predetermined maximum) that a squeeze has been attained, any still-pumpable cement slurry remaining in the drill pipe or tubing ideally can be reverse circulated out before it sets.

Usually the zone to be squeezed is isolated from above with a packer (and possibly from below with a bridge plug), but sometimes the squeezing pressure is applied to the entire casing string in what is known as a bradenhead squeeze, (named for an old manufacturer of casing heads). Even if a drilling rig is on location, pumping operations usually are done by a service company's cementing unit that can easily mix small batches of cement slurry, measure displacement volume accurately to spot the slurry on bottom, then pump at very low rates and high pressures during the squeeze itself, and finally measure volumes accurately again when reversing out any excess slurry. A squeeze manifold is a compact arrangement of valves and pressure gauges that allows monitoring of the drill pipe and casing pressures throughout the job, and facilitates quick switching of the pumping pressure to either side while the fluid returning from the other side of well is directed to the mud pit or a disposal pit or tank.

The generic term "squeeze" also can apply to injection of generally small volumes of other liquids (e.g., treating fluids) into a zone under pressure. Bullhead squeeze (or just plain bullheading) refers to pumping kill-weight mud down the casing beneath closed blowout preventers in a kick-control situation when it isn't feasible to circulate in such from bottom.

Differential Sticking

Differential sticking is a problem that occurs when drilling a well with a greater well bore pressure than formation pressure, as is usually the case. The drill pipe is pressed against the wellbore wall so that part of its circumference will see only reservoir pressure, while the rest will continue to be pushed by wellbore pressure. As a result the pipe becomes stuck to the wall, and can require millions of pounds of force to remove, which may prove impossible. In many cases the drilling fluid (mud) weight is simply reduced, thus relieving the pressure difference and releasing the stuck pipe string. Should this option be unavailable, as in sour gas wells, a specialty fishing company is called to retrieve the stuck pipe or 'fish'. Many options exist once a fishing company is on site: oil or nitrogen may be pumped down the well, or the fish may be 'washed over' using a carbide shoe on a string of washpipe. Jarring is not usually attempted with differential sticking due to the massive amount of pressure that holds the pipe in place.

Directional Drilling

Directional drilling (or slant drilling) is the practice of drilling non-vertical wells. It can be broken down into three main groups: Oilfield Directional Drilling, Utility Installation Directional Drilling (or H.D.D. or Horizontal Directional Drilling) Directional boring, and in-seam directional drilling to extract coal bed methane.

Many prerequisites enabled this suite of technologies to become productive. Probably, the first requirement was the realization that oil wells, or water wells, are not necessarily vertical. This realization was quite slow, and did not really grasp the attention of the oil industry until the late 1920s when there were several lawsuits alleging that wells drilled from a rig on one property had crossed the boundary and were penetrating a reservoir on an adjacent property. Initially, proxy evidence such as production changes in other wells was accepted, but such cases fuelled the development of small diameter tools capable of surveying wells during drilling.

Measuring the inclination of a wellbore (its deviation from the vertical) is comparatively simple, requiring only a pendulum. Measuring the azimuth (direction with respect to the geographic grid in which the wellbore is running from the vertical), however, was more difficult. In certain circumstances, magnetic fields could be used, but could be influenced by metalwork used inside wellbores, as well as the metalwork used in drilling equipment. The next advance was in the modification of small gyroscopic compasses by the Sperry Corporation, which was making similar compasses for aeronautical navigation. Sperry did this under contract to Sun Oil (which was involved in a lawsuit as described above), and a spin-off company "Sperry Sun" was formed, which brand continues to this day, absorbed into Halliburton.

Three components are measured at any given point in a wellbore in order to determine its position: the depth of the point along the course of the borehole (measured depth), the inclination at the point, and the magnetic azimuth at the point. These three components combined are referred to as a "survey". A series of consecutive surveys are needed to track the progress and location of a wellbore. Many of the earliest innovations such as photographic single shot technology and crow's feet baffle plates for landing survey tools were developed by Robert Richardson, an independent directional driller who first drilled in the 1940s and was still working in 2008.

Prior experience with rotary drilling had established several principles for the configuration of drilling equipment down hole

("Bottom Hole Assembly" or "BHA") that would be prone to "drilling crooked hole" (i.e., initial accidental deviations from the vertical would be increased). Counter-experience had also given early directional drillers ("DD's") principles of BHA design and drilling practice that would help bring a crooked hole nearer the vertical.

In 1934, H. John Eastman of Long Beach, California, became a pioneer in directional drilling when he and George Failing of Enid, Oklahoma, saved the Conroe, Texas, oil field. Failing had recently patented a portable drilling truck. He had started his company in 1931 when he mated a drilling rig to a truck and a power take-off assembly. The innovation allowed rapid drilling of a series of slanted wells. This capacity to quickly drill multiple relief wells and relieve the enormous gas pressure was critical to extinguishing the Conroe fire. (E&P, "Making a hole was hard work," Kris Wells, American Oil & Gas Historical Society Contributing Editor, Nov. 1, 2006 and "Technology and the Conroe Crater"). In a May, 1934, *Popular Science Monthly* article, it was stated that "Only a handfull of men in the world have the strange power to make a bit rotating a mile below ground at the end of a steel drill pipe, snake its way in a curve or around a dog leg angle, to reach a desired object." Eastman Whipstock, Inc., would become the world's largest directional company in 1973.

Combined, these survey tools and BHA designs made directional drilling possible, but it was perceived as arcane. The next major advance was in the 1970s, when downhole drilling motors (aka mud motors, driven by the hydraulic power of drilling mud circulated down the drill string) became common. These allowed the bit to be rotated on the bottom of the hole, while most of the drill pipe was held stationary. A piece of bent pipe (a "bent sub") between the stationary drill pipe and the top of the motor allowed the direction of the wellbore to be changed without needing to pull all the drill pipe out and place another whipstock. Coupled with the development of Measurement While Drilling tools (using mud pulse telemetry or EM telemetry, which allows tools down hole to send directional data back to the surface without disturbing drilling operations), directional drilling became easier. Certain profiles could not be drilled without the drill string rotating at all times.

Benefits

Wells are drilled directionally for several purposes:

- Increasing the exposed section length through the reservoir by drilling through the reservoir at an angle

- Drilling into the reservoir where vertical access is difficult or not possible. For instance an oilfield under a town, under a lake, or underneath a difficult to drill formation
- Allowing more wellheads to be grouped together on one surface location can allow fewer rig moves, less surface area disturbance, and make it easier and cheaper to complete and produce the wells. For instance, on an oil platform or jacket offshore, 40 or more wells can be grouped together. The wells will fan out from the platform into the reservoir(s) below. This concept is being applied to land wells, allowing multiple subsurface locations to be reached from one pad, reducing costs.
- Drilling along the underside of a reservoir-constraining fault allows multiple productive sands to be completed at the highest stratigraphic points.
- Drilling a "relief well" to relieve the pressure of a well producing without restraint (a "blow out"). In this scenario, another well could be drilled starting at a safe distance away from the blow out, but intersecting the troubled wellbore. Then, heavy fluid (kill fluid) is pumped into the relief wellbore to suppress the high pressure in the original wellbore causing the blowout.

Most directional drillers are given a well path to follow that is predetermined by engineers and geologists before the drilling commences. When the directional driller starts the drilling process, periodic surveys are taken with a downhole instrument to provide survey data (inclination and azimuth) of the well bore. These pictures are typically taken at intervals between 30–500 feet, with 90 feet common during active changes of angle or direction, and distances of 200–300 feet being typical while "drilling ahead" (not making active changes to angle and direction). During critical angle and direction changes, especially while using a downhole motor, an MWD (Measurement while drilling) tool will be added to the drill string to provide continuously updated measurements that may be used for (near) real-time adjustments.

These data indicate if the well is following the planned path and whether the orientation of the drilling assembly is causing the well to deviate as planned. Corrections are regularly made by techniques as simple as adjusting rotation speed or the drill string weight (weight on bottom) and stiffness, as well as more complicated and time consuming methods, such as introducing a downhole motor. Such pictures, or surveys, are plotted and maintained as an engineering and legal record describing the path of the well bore. The survey pictures taken while drilling are typically confirmed by a later survey in full of the borehole, typically using a "multi-shot camera" device.

The multi-shot camera advances the film at time intervals so that by dropping the camera instrument in a sealed tubular housing inside the drilling string (down to just above the drilling bit) and then withdrawing the drill string at time intervals, the well may be fully surveyed at regular depth intervals (approximately every 90 feet being common, the typical length of 2 or 3 joints of drill pipe, known as a stand, since most drilling rigs "stand back" the pipe withdrawn from the hole at such increments, known as "stands").

Drilling to targets far laterally from the surface location requires careful planning and design. The current record holders manage wells over 10 km (6.2 mi) away from the surface location at a true vertical depth (TVD) of only 1,600–2,600 m (5,200–8,500 ft).

Disadvantages

Until the arrival of modern downhole motors and better tools to measure inclination and azimuth of the hole, directional drilling and horizontal drilling was much slower than vertical drilling due to the need to stop regularly and take time consuming surveys, and due to slower progress in drilling itself (lower rate of penetration). These disadvantages have shrunk over time as downhole motors became more efficient and semi-continuous surveying became possible.

What remains is a difference in operating costs: for wells with an inclination of less than 40 degrees, tools to carry out adjustments or repair work can be lowered by gravity on cable into the hole. For higher inclinations, more expensive equipment has to be mobilized to push tools down the hole. Another disadvantage of wells with a high inclination was that prevention of sand influx into the well was less reliable and needed higher effort. Again, this disadvantage has diminished such that, provided sand control is adequately planned, it is possible to carry it out reliably.

Stealing Oil

In 1990, Iraq accused Kuwait of stealing Iraq's oil through slant drilling. The United Nations redrew the border after the 1991 Gulf war that liberated Kuwait from a seven-month Iraqi occupation under former leader Saddam Hussein. It placed 11 oil wells, some farms and an old naval base that used to be in Iraq on the Kuwaiti side.

In the mid-twentieth century, a slant-drilling scandal occurred in the huge East Texas Oil Field.

Geosteering

In the process of drilling a borehole, geosteering is the act of adjusting the borehole position (inclination and azimuth angles) *on the fly* to reach one or more geological targets. These changes are based on geological information gathered while drilling.

Description

From 2D and 3D models of underground substructures, deviated wells (2D and 3D) are planned in advance to achieve specific goals: exploration, fluids production, fluids injection or technical.

A *well plan* is a continuous succession of straight and curved lines representing the geometrical figure of the expected well path. A well plan is always projected on vertical and horizontal maps.

While the borehole is being drilled according to the well plan, new geological information is gathered from mud logging, Measurement While Drilling and Logging While Drilling. These usually show some differences from what is expected from the model. As the model is continuously updated with the new geological information (formation evaluation) and the borehole position (*well deviation survey*), changes start to appear in the geological substructures and can lead to the well plan being updated to reach the corrected geological targets.

Drilling Engineering

Drilling engineering is a subset of petroleum engineering. Drilling engineers design and implement procedures to drill wells as safely and economically as possible. They work closely with the drilling contractor, service contractors, and compliance personnel, as well as with geologists and other technical specialists. The drilling engineer has the responsibility for ensuring that costs are minimized while getting information to evaluate the formations penetrated, protecting the health and safety of workers and other personnel, and protecting the environment.

The planning phases involved in drilling an oil or gas well typically involve estimating the value of sought reserves, estimating the costs to access reserves, acquiring property by a mineral lease, a geological survey, a well bore plan, and a layout of the type of equipment required to reach the depth of the well. Drilling engineers are in charge of the process of planning and drilling oil wells. Their responsibilities include:

1. Designing well programs (e.g., casing sizes and setting depths) to prevent blowouts (uncontrolled well-fluid release) while allowing adequate formation evaluation.

2. Designing or contributing to the design of casing strings and cementing plans, directional drilling plans, drilling fluids programs, and drill string and drill bit programs.
3. Specifying equipment, material and ratings and grades to be used in the drilling process.
4. Providing technical support and audit during the drilling process.
5. Performing cost estimates and analysis.
6. Developing contracts with vendors.

Drilling engineers are often degreed as petroleum engineers, although they may come from other technical disciplines (i.e., mechanical engineering or geology) and subsequently be trained by an oil and gas company. They also may have practical experience as a rig hand or mudlogger or mud engineer.

"Mud engineer" is the name given to an oil field service company individual who is charged with maintaining a drilling fluid or completion fluid system on an oil and/or gas drilling rig. This individual typically works for the company selling the chemicals for the job and is specifically trained with those products, though independent mud engineers are still common. The work schedule of the mud engineer or more properly Drilling Fluids Engineer, is arduous, often involving long shifts. Typical shifts range from twenty-one days on and seven off, to five weeks on and five weeks off. Consulting mud men can work up to one hundred days or longer, in a row. Nowadays, the internationally accepted shift pattern at off-shore drilling operations is personnel (including mud engineers) work on a 28 day shift pattern, where they work for 28 continuous days and rest the following 28 days.

In offshore drilling, with new technology and high total day costs, wells are being drilled extremely fast. Having two mud engineers makes economic sense to prevent down time due to drilling fluid difficulties. Two mud engineers also reduce insurance costs to oil companies for environmental damage that oil companies are responsible for during drilling and production. A senior mud engineer typically works in the day, and a junior mud engineer at night.

The cost of the drilling fluid is typically about 10% (may vary greatly) of the total cost of drilling a well, and demands competent mud engineers. Large cost savings result when the mud engineer and fluid performs adequately. The mud engineer is not to be confused with mudloggers, service personnel who monitor gas from the mud and collect well bore samples.

Compliance Engineer

The compliance engineer is the most common name for a relatively new position in the oil field, emerging around 2002 due to new environmental regulations on synthetic mud in the United States. Previously, synthetic mud was regulated the same as water-based mud and could be disposed of in offshore waters due to low toxicity to marine organisms. New regulations restrict the amount of synthetic oil that can be discharged. These new regulations created a significant burden in the form of tests needed to determine the "ROC" or retention on cuttings, sampling to determine the percentage of crude oil in the drilling mud, and extensive documentation. It should be noted that no type of oil/synthetic based mud (or drilled cuttings contaminated with OBM/SBM) may be dumped in the North Sea. Contaminated mud must either be shipped back to shore in skips or processed on the rigs.

A new monthly toxicity test is also now performed to determine sediment toxicity, using the amphipod *Leptocheirus plumulosus*. Various concentrations of the drilling mud are added to the environment of captive *L. plumulosus* to determine its effect on the animals. The test is controversial for two reasons:

1. These animals are not native to many of the areas regulated by them, including the Gulf of Mexico
2. The test has a very large standard deviation and samples that fail badly may pass easily upon retesting

Mud Engineer

A mud engineer (correctly called a Drilling Fluids Engineer, but most often referred to as the "Mud Man") works on an oil well or gas well drilling rig, and is responsible ensuring the properties of the drilling fluid, also known as drilling mud, are within designed specifications.

Use of Mud

Mud is a vital part of drilling operations. It provides hydrostatic pressure on the borehole wall to prevent uncontrolled production of reservoir fluids, lubricates and cools the drill bit, carries the drill cuttings up to the surface and forms a "filter-cake" on the borehole wall to prevent drilling fluid invasion. To fulfil these tasks effectively, the mud contains carefully chosen additives to control its chemical and rheological properties.

Drilling mud is usually a shear-thinning non-Newtonian fluid of variable viscosity. When it is under more shear, such as in the pipe to the bit and through the bit nozzles, viscosity is lower which reduces pumping-power requirements. When returning to the surface through the much roomier annulus it is under less shear stress and becomes more viscous, and hence better able to carry the rock cuttings. Bentonite is commonly used as an additive to control and maintain viscosity, and also has the additional benefit of forming a mud-cake (also known as a filter cake) on the bore-hole wall, preventing fluid invasion.

Barite is commonly used to "weight" the mud to maintain adequate hydrostatic pressure down-hole. This is critical in a drilling operation to avoid a kick and ultimately a blowout from uncontrolled production of formation fluids. The "mud-pits" at the surface have their levels carefully monitored, since an increase in the mud level indicates a kick is taking place, and may require shutting in the well and circulating heavier weighted drilling mud to prevent further formation fluid or gas production. Drilling fluid must be chemically compatible with the formations being drilled. Salinity must be chosen so as not to cause clay swelling or other problems. Mud can be "oil-based" or "water-based". In many areas oil-based muds are being phased out, as they are less environmentally friendly, although in some formations they are necessary because of chemical compatibility issues. Offshore rigs typically use synthetic oil based mud.

The Job

The mud engineer (or drilling fluids engineer) may be a university, college, or technical institute graduate, or may have no tertiary education at all, having gained experience working on rigs which could be over 10 years. On land, this experience would come from being a derrick hand, and offshore, the experience would come from being a pump man. Prior to working on his own, he has been on a special training course, known as "mud school", and often spends time working with a senior mud engineer to gain experience. Prior to drilling a well, a "mud program" will be worked out according to the expected geology, in which products to be used, concentrations of those products, and fluid specifications at different depths are all predetermined. As the hole is drilled and gets deeper, more mud is required, and the mud engineer is responsible for making sure that the new mud to be added is made up to the required specifications. The chemical composition of the mud will be designed so as to stabilize the hole. It is sometimes necessary to completely change the mud to drill through a particular subsurface layer.

As drilling proceeds, the mud engineer will get information from the mud logger (mud logging technician) about progress through the geological zones, and will make regular physical and chemical checks on the drilling mud. In particular the Marsh funnel viscosity and the density are frequently checked. As drilling proceeds, the mud tends to accumulate small particles of the rocks which are being drilled through, and its properties change. It is the job of the mud engineer to specify additives to correct these changes, or to partially or wholly replace the mud when necessary. He or she must also keep an eye on the equipment which is used to pump the mud and to remove particles, and be prepared if the geologists' predictions are not entirely correct, or if other problems arise. It is sometimes necessary to stabilize the wall of a borehole at a particular depth by pumping cement down through the mud system, and the mud engineer is sometimes in charge of this process. The mud engineer is well supported by the mud supply company with computer aids and manuals dealing with all known problems and their solution, but it is his or her responsibility to get it right in a situation where mistakes can be very costly indeed.

A mud engineer's job may involve long shifts of over 12 hours a day. Typical offshore and foreign work schedules are four weeks working and four weeks off.

Important Fluid Properties

One of the most important mud properties is the mud weight (density). If the mud weight exceeds the fracture pressure of the formation, the formation may fracture and large quantities of mud are lost to it, in a situation referred to as lost circulation. These cracks can also cause water to seep into the well bore or into a hydrocarbon bearing zone, which would likely impede the ability of the formation to produce oil (or require the separation of large quantities of water).

Conversely, if the mud weight is too low it will have a hydrostatic pressure that is less than the formation pressure. This will cause pressurized fluid in the formation to flow into the wellbore and make its way to the surface. This is referred to as a formation "kick" and can lead to a potentially deadly blowout if the invading fluid reaches the surface uncontrolled. Other important mud properties to be maintained are the YP (Yield Point) which determines the carrying capacity of the mud to carry the drill cuttings to the surface. Mud should be capable of forming a thin "mud cake" which forms a lining of the borehole walls.

Drilling Fluids Companies

Drilling fluids operations are often contracted to service companies, a trend commonly observed in the oil industry for most of it operations. The largest three companies for mud services are M-I SWACO (A Schlumberger Company), Baroid Drilling Fluids (Halliburton Oilfield

Services), and Baker Hughes Drilling Fluids. There are, however, many smaller companies providing drilling fluid services as well.

Drilling Fluid Invasion

Drilling fluid invasion is a process that occurs in a well being drilled with higher wellbore pressure than formation pressure. The liquid component of the drilling fluid (known as the mud filtrate) continues to "invade" the porous and permeable formation until the solids present in the mud, commonly bentonite, clog enough pores to form a mud cake capable of preventing further invasion.

If invasion is severe enough, and reservoir pressures are unable to force the fluid and associated particles out entirely when the well starts producing, the amount of oil and gas a well can produce can be permanently reduced. This is especially true when a process called phase trapping occurs. This is when a fluid enters a formation that is below its irreducible saturation of that fluid. Once the fluid is present, it is held in place by capillary forces and usually can not be removed.

Invasion also has significant implications for well logging. In many cases the "depth of investigation" of a well logging tool is only a few inches (or even less for things like sonic logs), and it is quite possible that drilling fluid has invaded beyond this depth. Therefore the readings that are taken that are influenced by formation fluids are measuring mud filtrate properties rather than formation (in situ) fluid properties.

Drill Stem Test

A drill stem test (DST) is a procedure for isolating and testing the surrounding geological formation through the drill stem. The test is a measurement of pressure behaviour at the drill stem and is a valuable way to obtain important sampling information on the formation fluid and to establish the probability of commercial production.

In oil and natural gas extraction, the drill stem includes the drill pipe, drill collars, bottomhole assembly, and drill bit. During normal drilling, fluid is pumped through the drill stem and out the drill bit. In a drill stem test, the drill bit is removed, a drill stem test tool is added, and fluid from the formation is recovered through the drill stem, while several measurements of pressure are being made.

The basic drill stem test tool consists of a packer or packers, valves or ports that may be opened and closed from the surface, and two or more pressure-recording devices. (A packer is an expanding plug which can be used to seal off sections of the open or cased well, to isolate them for testing.) The drill stem test tool is lowered on the drill pipe to the zone to be tested. The packer or packers are set to isolate the zone from the drilling fluid column, the tester valve is opened, and testing begins.

History

Working in El Dorado, Arkansas, in the 1920s, E.C. Johnston and his brother M.O. Johnston developed the first drill stem tester and ran the first commercial drill stem test in 1926. In April 1929, the Johnston Formation Testing Corporation was granted a patent (U.S. Patent 1,709,940) and they subsequently refined the testing system in the early 1930s.

In the 1950s, Schlumberger introduced a method for testing formations using wireline. The Schlumberger formation-testing tool, placed in operation in 1953, fired a shaped charge through a rubber pad that had been expanded in the hole until it was securely fixed in the hole at the depth required. Formation fluids flowed through the perforation and connecting tubing into a container housed inside the tool. When filled, the container was closed, sealing the fluid sample at the formation pressure. The tool was then brought to the surface, where the sample could be examined. In 1956, Schlumberger acquired Johnston Testers and continues to perform drill stem tests and wireline formation tests in both open and cased holes.

Lost Circulation

Lost circulation can be one of the more serious problems that can arise during the drilling of an oil well or gas well. Circulation is said to be lost when the drilling fluid, known commonly as "mud", flows into one or more geological formations instead of returning up the annulus.

Consequences

The consequences of lost circulation can be as little as the loss of a few dollars of drilling fluid, or as disastrous as a blowout and loss of life, so close monitoring of tanks, pits, and flow from the well, to quickly assess and control lost circulation, is taught and practiced. If the fluid in the wellbore drops due to lost circulation (or any other reason), hydrostatic pressure is reduced, thus allowing a gas or fluid, which is under a higher pressure than the reduced hydrostatic pressure, to influx into the wellbore.

Another consequence of lost circulation is called "dry drilling". Dry drilling occurs when fluid is completely lost from the well bore without actual drilling coming to a stop. The effects of dry drilling can be as minor as destroying a bit to as serious as major damage to the wellbore requiring a new well to be drilled. Dry drilling can also cause severed damage to the drill string, including snapping the pipe, and the drilling rig itself.

Categories

Lost circulation falls into two main categories:

- "Minor losses" - This is where losses are between 6 and 470 barrels (1 to 75 m^3), and remains within those amounts, or is ceased, within 48 hours.
- "Severe losses" - This is where losses are greater than 470 barrels (75 m^3), or it takes greater than 48 hours to control or cease the lost circulation.

Total losses may also occur, where the return of fluids is completely lost to the surface. This may fall into either the minor or severe losses categories, depending on the amount of losses and the time involved in regaining circulation.

Control

Although preferred, ceasing lost circulation completely is not always possible or required. Controlled losses allows drilling to continue while keeping the wellbore full, preventing an influx of gas or fluid into the wellbore, known as a" kick", which can lead to a blowout.

A number of options are available when lost circulation occurs, depending on the severity.

- Minor losses may be controlled by increasing the viscosity of the fluid with bentonite and/or polymers, or with the addition of other additives which typically includes sawdust.
- Severe losses will require increasing the viscosity of the fluid with bentonite and/or polymers and the addition of other additives which typically includes sawdust.

Total losses can be regained through conventional use of increased viscosity and additives, or through use of unconventional methods such as pumping of golf balls, tree branches, rags, additive sacks, and other items in conjunction with, or followed by, a high viscosity fluid.

If total losses occur and circulation cannot be regained, several options are available, depending on the operational requirements and depth being drilled in relation to desired production geological zones. Continuing drilling while pumping drilling fluid is one option, though continued drilling while pumping water is less costly and more often used.

Sometimes the cuttings from continued drilling will aid in reducing or stop losses altogether. A third option is to cement the zone where the losses occurred, and to drill through the cement and continue drilling the well. This third option is very often the most cost effective if severe losses occur, as lost circulation can sometimes not be controlled with conventional or unconventional methods.

Additives

- The most common additive used to control or cease lost circulation is bentonite, in that it inherently will seal small holes or fractures. Bentonite, in higher concentrations, increases viscosity, and therefore slows the fluid flow into the surrounding rock. Although bentonite is the most common additive used, it sometimes is not used at all, depending on the current fluid being used and the current depth of drilling in relation to desired production zones.
- Polymers are also sometimes used to increase the viscosity. Though these are more costly, they are more compatible with several types of fluid systems.
- When it comes to additives which physically plug or seal the losses, there are a great deal of them. Sawdust, flaked cellophane, crushed or ground gypsum are all used.
- Other common, and cheaper additives, that are used are shredded newspaper and cotton seed hulls. Cotton seed hulls are less preferred as they may cause wear to pump swabs and springs. Both of these are generally only used when either fresh or brine water is being used for the drilling fluid.
- When drilling in salt formations, brine water is typically used as it is more difficult to dissolve more of the salt preventing the formation of "washouts". Washouts not only contribute to loss of circulation, but can jeopardize the integrity of the wellbore itself.

Additive Considerations

Several factors are considered in what additives are used:

- Hole size currently being drilled.
- Drilling fluid in use. The additives must be compatible.
- Depth of the well in regards to geological stability.
- Depth of well in relation to the desired production zones. Plugging a production zone is not a desired outcome.

- Drill bit nozzles sizes. If the additive(s) will not go through the drill bit, they cannot be used.
- Other drill string mechanical equipment such as a mud motor or MWD tools. If the additive(s) will not go through the drill string, they cannot be used.

Tracer use in the Oil Industry

Tracers are used in the oil industry in order to qualitatively or quantitatively gauge how fluid flows through the reservoir, as well as being a useful tool for estimating residual oil saturation. Tracers can be used in either interwell tests or single well tests. In interwell tests, the tracer is injected at one well along with the carrier fluid (water in a waterflood or gas in a gasflood) and detected at a producing well after some period of time, which can be anything from days to years.

In single well tests, tracer is injected into the formation from a well and then produced out the same well. The delay between a tracer that does not react with the formation (a conservative tracer) and one that does (a partitioning tracer) will give an indication of residual oil saturation, a piece of information that is difficult to acquire by other means.

Tracers can be radioactive or chemical, gas or liquid and have been used extensively in the oil industry and hydrology for decades.

Underbalanced Drilling

Underbalanced drilling, or UBD, is a procedure used to drill oil and gas wells where the pressure in the wellbore is kept lower than the fluid pressure in the formation being drilled. As the well is being drilled, formation fluid flows into the wellbore and up to the surface. This is the opposite of the usual situation, where the wellbore is kept at a pressure above the formation to prevent formation fluid entering the well. In such a conventional "overbalanced" well, the invasion of fluid is considered a kick, and if the well is not shut-in it can lead to a blowout, a dangerous situation. In underbalanced drilling, however, there is a "rotating head" at the surface - essentially a seal that diverts produced fluids to a separator while allowing the drill string to continue rotating.

If the formation pressure is relatively high, using a lower density mud will reduce the well bore pressure below the pore pressure of the formation. Sometimes an inert gas is injected into the drilling mud to reduce its equivalent density and hence its hydrostatic pressure throughout the well depth. This gas is commonly nitrogen, as it is noncombustible and readily available, but air, reduced oxygen air, processed flue gas and natural gas have all been used in this fashion.

Kinds of Underbalanced Drilling

There are several kinds of underbalanced drilling. The most common are listed below.

- Dry Air. This is also known as dusting. Here air compressors combined with a booster (which takes the head from the compressors and increases the pressure of the air, but does not increase the volume of air going down hole) are used and the only fluid injected into the well is a small amount of oil to reduce corrosion.
- Mist. A small amount of foaming agent (soap) is added into the flow of air. Fine particles of water and foam in an atmosphere of air bring cuttings back to the surface.
- Foam. A larger amount of foaming agent is added into the flow. Bubbles and slugs of bubbles in an atmosphere of mist bring cuttings back to the surface.
- Stable foam. An even larger amount of foaming agent is added into the flow. This is the consistency of a shaving cream.
- Airlift. Slugs and bubbles of air in a matrix of water, soap can or can not be added into the fluid flow of air.
- Aerated Mud. Air or another gas is injected into the flow of drilling mud. Degassing units are required to remove air before it can be recirculated.

Advantages

Underbalanced wells have several advantages over conventional drilling including:

- Eliminated formation damage. In a conventional well, drilling mud is forced into the formation in a process called invasion, which frequently causes formation damage - a decrease in the ability of the formation to transmit oil into the wellbore at a given pressure and flow rate. It may or may not be repairable. In underbalanced drilling, if the underbalanced state is maintained until the well becomes productive, invasion does not occur and formation damage can be completely avoided.
- Increased Rate of Penetration (ROP). With less pressure at the bottom of the wellbore, it is easier for the drill bit to cut and remove rock.

- Reduction of lost circulation. Lost circulation is when drilling mud flows into the formation uncontrollably. Large amounts of mud can be lost before a proper mud cake forms, or the loss can continue indefinitely. If the well is drilled underbalanced, mud will not enter the formation and the problem can be avoided.
- Differential sticking is eliminated. Differential sticking is when the drill pipe is pressed against the wellbore wall so that part of its circumference will see only reservoir pressure, while the rest will continue to be pushed by wellbore pressure. As a result the pipe becomes stuck to the wall, and can require thousands of pounds of force to remove, which may prove impossible. Because the reservoir pressure is greater than the wellbore pressure in UBD, the pipe is pushed away from the walls, eliminating differential sticking.
- Formation damage Some rock formation have a reactive tendency to water. When drillmud is used the water in the drill mud reacts with the formation (mostly clay) and inheriently causes a formation damage (reduction in permeability and porosity) Use of underbalanced drilling can prevent it

Disadvantages

Underbalanced drilling is usually more expensive than conventional drilling (when drilling a deviated well which requires directional drilling tools), and has safety issues of its own. Technically the well is always in a blowout condition unless a heavier fluid is displaced into the well. Air drilling requires a faster up hole volume as the cuttings will fall faster down the annulus when the compressors are taken off the hole compared to having a higher viscosity fluid in the hole. Because air is compressible mud pulse telemetry measurement while drilling (MWD) tools which require an incompressible fluid can not work. Common technologies used to eliminate this problem are either electromagnetic MWD tools or wireline MWD tools. Downhole mechanics are usually more violent also because the volume of fluid going through a downhole motor or downhole hammer is greater than an equivalent fluid when drilling balanced or over balanced because of the need of higher up hole velocities. Corrosion is also a problem, but can be largely avoided using a coating oil or rust inhibitors.

Well Logging

Well logging, also known as borehole logging is the practice of making a detailed record (a *well log*) of the geologic formations penetrated by a borehole. The log may be based either on visual inspection of samples brought to the surface (*geological* logs) or on physical measurements made by instruments lowered into the hole (*geophysical* logs). Well logging is done during all phases of a well's development; drilling, completing, producing and abandoning. Mostly in the oil and gas, groundwater, minerals, geothermal, and for environmental and geotechnical studies.

Electric or Geophysical Well Logs

The oil and gas industry records rock and fluid properties to find hydrocarbon zones in the geological formations intersected by a borehole. The logging procedure consists of lowering a 'logging tool' on the end of a wireline into an oil well (or hole) to measure the rock and fluid properties of the formation. An interpretation of these measurements is then made to locate and quantify potential depth zones containing oil and gas (hydrocarbons). Logging tools developed over the years measure the electrical, acoustic, radioactive, electromagnetic, nuclear magnetic resonance, and other properties of the rocks and their contained fluids. Logging is usually performed as the logging tools are pulled out of the hole. This data is recorded either at surface (real-time mode), or downhole (memory mode) to electronic data format and then either a printed record or electronic presentation called a "well log" provided to the client. Well logging is performed at various intervals during the drilling of the well and when the total depth is drilled, which could range in depths from 150 m to 10668 m (500 ft to 35,000 ft) or more.

Electric line is the common term for the armoured, insulated cable used to conduct current to downhole tools used for well logging. Electric line can be subdivided into open hole operations and cased hole operations. Other conveyance methods for logging are logging while drilling (LWD), tractor, coiled tubing (real-time and memory), drill pipe conveyed, and slickline (memory, and with new development, some slickline telemetry capability).

Open hole operations, or reservoir evaluation, involve the deployment of tools into a freshly drilled well. As the toolstring traverses the wellbore, the individual tools gather information about the surrounding formations. A typical open hole log will have information about the density, porosity, permeability, lithology, presence of hydrocarbons, and oil and water saturation.

Cased hole operations, or production optimization, focus on optimizing the completed oil well through mechanical services and logging technologies. At this point in the well's life, the well is encased in steel pipe, cemented into the well bore and may or may not be producing. A typical cased hole log may show cement quality, production information, and formation data. Mechanical services use jet perforating guns, setting tools, and dump bailors to optimize the flow of hydrocarbons.

Wireline Tool Types

Typically the wireline tools are cylindrical in shape, usually from 1.5 to 5 inches in diameter. "Open hole" tool combinations can extend to over 100 feet long; "cased hole" tool combinations are often limited in length by the height restrictions imposed by constraints of "lubricator" pipe section required to contain the well pressure while deploying cased hole tools. There are many types of logging tools, ranging from common measurements (pressure and temperature) to advanced rock properties and fracture analysis, fluid properties in the wellbore, or formation properties extending several meters into the rock formation.

1. With sensors without excitation: There are units to measure spontaneous potential (SP), which is a voltage difference between a surface electrode and another electrode located in the downhole instrument, other instruments that measure the natural radiation from natural isotopes of potassium, thorium, etc., to measure pressure and temperature, etc.
2. With sources of excitation and sensors: There are sensor systems consistent with a source of excitation and a sensor. In this type we find acoustic (also called sonic), electric, inductive, magnetic resonance, sensing systems, just to name a few.
3. Instruments that produce some mechanical work, or retrieve a sample of fluid or rock to the surface.

Devices to collect samples of rock, samples of fluid extracted from the rock, and some other mechanical devices.

Types of Electric/Electronic Logs

There are many types of electric/electronic logs and they can be categorized either by their function or by the technology that they use. "Open hole logs" are run before the oil or gas well is lined with pipe or cased. "Cased hole logs" are run after the well is lined with casing or production pipe.

Electric/electronic logs can also be divided into two general types based on what physical properties they measure. Resistivity logs measure some aspect of the specific resistance of the geologic formation. There are about 17 types of resistivity logs.

Porosity logs measure the fraction or percentage of pore volume in a volume of rock. Most porosity logs use either acoustic or nuclear technology. Acoustic logs measure characteristics of sound waves propagated through the well-bore environment. Nuclear logs utilize nuclear reactions that take place in the downhole logging instrument or in the formation. Nuclear logs include density logs and neutron logs, as well as gamma ray logs which are used for correlation.

The basic principle behind the use of nuclear technology is that a neutron source placed near the formation of which the porosity is required to be measured will result in neutrons being scattered by the hydrogen atoms, largely those present in the formation fluid. Since there is little difference in the neutrons scattered by hydrocarbons or water, the porosity measured gives a figure close to the true physical porosity whereas the figure obtained from electrical resistivity measurements is that due to the conductive formation fluid.

The difference between neutron porosity and electrical porosity measurements therefore indicates the presence of hydrocarbons in the formation fluid.

History

Conrad and Marcel Schlumberger, who founded Schlumberger Limited in 1926, are considered the inventors of electric well logging. Conrad developed the Schlumberger array, which was a technique for prospecting for metal ore deposits, and the brothers adopted that surface technique to subsurface applications.

On September 5, 1927, a crew working for Schlumberger lowered an electric sonde or tool down a well in Pechelbronn, Alsace, France creating the first well log. In modern terms, the first log was a resistivity log that could be described as 3.5-meter upside-down lateral log. In 1931, Henri George Doll and G. Dechatre, working for Schlumberger, discovered that the galvanometer wiggled even when no current was being passed through the logging cables down in the well. This led to the discovery of the spontaneous potential (SP) which was as important as the ability to measure resistivity.

The SP effect was produced naturally by the borehole mud at the boundaries of permeable beds. By simultaneously recording SP and resistivity, loggers could distinguish between permeable oil-bearing beds and impermeable nonproducing beds.

In 1940, Schlumberger invented the spontaneous potential dipmeter; this instrument allowed the calculation of the dip and direction of the dip of a layer. The basic dipmeter was later enhanced by the resistivity dipmeter (1947) and the continuous resistivity dipmeter (1952).

Oil-based mud (OBM) was first used in Rangely Field, Colorado in 1948. Normal electric logs require a conductive or water-based mud, but OBMs are nonconductive. The solution to this problem was the induction log, developed in the late 1940s.

The introduction of the transistor and integrated circuits in the 1960s made electric logs vastly more reliable. Computerization allowed much faster log processing, and dramatically expanded log data-gathering capacity. The 1970s brought more logs and computers. These included combo type logs where resistivity logs and porosity logs were recorded in one pass in the borehole.

The two types of porosity logs (acoustic logs and nuclear logs) date originally from the 1940s. Sonic logs grew out of technology developed during World War II. Nuclear logging has supplemented acoustic logging, but acoustic or sonic logs are still run on some combination logging tools.

Nuclear logging was initially developed to measure the natural gamma radiation emitted by underground formations. However, the industry quickly moved to logs that actively bombard rocks with nuclear particles. The gamma ray log, measuring the natural radioactivity, was introduced by Well Surveys Inc. in 1939, and the WSI neutron log came in 1941. The gamma ray log is particularly useful as shale beds which often provide a relatively low permeability cap over hydrocarbon reservoirs usually display a higher level of gamma radiation. These logs were important because they can be used in cased wells (wells with production casing). WSI quickly became part of Lane-Wells. During World War II, the US Government gave a near wartime monopoly on open-hole logging to Schlumberger, and a monopoly on cased-hole logging to Lane-Wells. Nuclear logs continued to evolve after the war.

The nuclear magnetic resonance log was developed in 1958 by Borg Warner. Initially the NMR log was a scientific success but an engineering failure. However, the development of a continuous NMR logging tool by Numar (now a subsidiary of Halliburton) is a promising new technology.

Many modern oil and gas wells are drilled directionally. At first, loggers had to run their tools somehow attached to the drill pipe if the well was not vertical. Modern techniques now permit continuous information at the surface. This is known as logging while drilling (LWD) or measurement-while-drilling (MWD). MWD logs use mud pulse technology to transmit data from the tools on the bottom of the drillstring to the processors at the surface.

Logging While Drilling

In the 1970s, a new technique, logging while drilling (LWD), was introduced which provided similar information about the well. Instead of sensors being lowered into the well at the end of wireline cable, the sensors are integrated into the drill string and the measurements are made while the well is being drilled. While wireline well logging occurs after the drill string is removed from the well, LWD measures and transmits geological parameters while the well is being drilled. Measured data is transmitted to the surface in real time via pressure pulses in the well's mud fluid column.

This mud telemetry method provides a bandwidth of less than 10 bits per second, although, as drilling through rock is a fairly slow process, data compression techniques mean that this is an ample bandwidth for real-time delivery of information. A higher sample rate of data is recorded into memory and retrieved when the drillstring is withdrawn at bit changes.

Logging Measurement Types

Logging measurements are quite sophisticated. The prime target is the measurement of various geophysical properties of the subsurface rock formations. Of particular interest are porosity, permeability, and fluid content. Porosity is the proportion of fluid-filled space found within the rock. It is this space that contains the oil and gas. Permeability is the ability of fluids to flow through the rock. The higher the porosity, the higher the possible oil and gas content of a rock reservoir. The higher the permeability, the easier for the oil and gas to flow toward the wellbore. Logging tools provide measurements that allow for the mathematical interpretation of these quantities.

Beyond just the porosity and permeability, various logging measurements allow the interpretation of what kinds of fluids are in the pores—oil, gas, brine. In addition, the logging measurements are used to determine mechanical properties of the formations. These mechanical properties determine what kind of enhanced recovery methods may be used (tertiary recovery) and what damage to the formation (such as erosion) is to be expected during oil and gas production.

The types of instruments used in well logging are quite broad. The first logging measurements consisted of basic electrical resistivity logs and spontaneous potential (SP) logs, introduced by the Schlumberger brothers in the 1920s. Tools later became available to estimate porosity via sonic velocity and nuclear measurements. Tools are now more specialized and better able to resolve fine details in the formation. Radiofrequency transmission and coupling techniques are used to determine electrical conductivity of fluid (brine is more conductive than oil or gas). Sonic transmission characteristics (pressure waves) determine mechanical integrity.

Nuclear magnetic resonance (NMR) can determine the properties of the hydrogen atoms in the pores (surface tension, etc.). Nuclear scattering (radiation scattering), spectrometry and absorption measurements can determine density and elemental analysis or composition. High resolution electrical or acoustical imaging logs are used to visualize the formation, compute formation dip, and analyze thinly-bedded and fractured reservoirs.

In addition to sensor-based measurements above, robotic equipment can sample formation fluids which may then be brought to the surface for laboratory examination. Also, controlled flow measurements can be used to determine in situ viscosity, water and gas cut (percentage), and other fluid and production parameters.

Geological Logs

Geological logs use data collected at the surface, rather than by downhole instruments. The geological logs include *drilling time logs*, *core logs*, *sample logs*, and *mud logs*. Mud logs have become the oil industry standard.

Drilling time logs record the time required to drill a given thickness of rock formation. A change in the drilling rate or penetration rate usually means a change in the type of rock penetrated by the bit. The drilling time is expressed as minutes per foot, while the rate of penetration is usually expressed as feet per hour. Therefore, drilling time is the inverse of penetration rate.

Sample logs are made by examining cuttings, which are bits of rock circulated to the surface by the drilling mud in rotary drilling. The cuttings have travelled up the wellbore suspended in the drilling fluid or mud which was pumped into the wellbore via the drill string/ pipe and they return to the surface via the annulus, then to the shale shakers via the flow line. Cuttings are then separated from the drilling fluid as they move across the shale shakers and are sampled at regular depth intervals. These rock samples are analyzed and described by the wellsite geologist or mudlogger.

Mud logs are prepared by a mud logging company contracted by the operating company. One parameter a typical mud log displays is the formation gas (gas units or ppm). "The gas recorder usually is scaled in terms of arbitrary gas units, which are defined differently by the various gas-detector manufactures. In practice, significance is placed only on relative changes in the gas concentrations detected." The current industry standard mud log normally includes real-time drilling parameters such as rate of penetration (ROP), lithology, gas hydrocarbons, flow line temperature (temperature of the drilling fluid) and chlorides but may also include mud weight, estimated pore pressure and corrected d-exponent (corrected drilling exponent) for a pressure pack log. Other information that is normally notated on a mud log include lithology descriptions, directional data (deviation surveys), weight on bit, rotary speed, pump pressure, pump rate, viscosity, drill bit info, casing shoe depths, formation tops, mud pump info, to name just a few.

Wireline log

A continuous measurement of formation properties with electrically powered instruments to infer properties and make decisions about drilling and production operations. The record of the measurements, typically a long strip of paper, is also called a log. Measurements include electrical properties (resistivity at various frequencies), sonic properties, active and passive nuclear measurements, dimensional measurements of the wellbore, formation fluid sampling, formation pressure measurement, wireline-conveyed sidewall coring tools, and others.

In wireline measurements, the logging tool (or probe) is lowered into the open wellbore on a multiple conductor, contra-helically armoured wireline. Once lowered to the bottom of the interval of interest, the measurements are taken on the way out of the wellbore. This is done in an attempt to maintain tension on the cable (which stretches) as constant as possible for depth correlation purposes. (The exception to this practice is in certain hostile environments in which the tool electronics might not survive the temperatures on bottom for the amount of time it takes to lower the tool and then record measurements while pulling the tool up the hole.

In this case, "down log" measurements might actually be conducted on the way into the well, and repeated on the way out if possible.) Most wireline measurements are recorded continuously even though the probe is moving. Certain fluid sampling and pressure-measuring tools require that the probe be stopped, increasing the chance that the probe or the cable might become stuck. LWD tools take measurements in much the same way as wireline-logging tools, except that the measurements are taken by a self-contained tool near the bottom of the bottomhole assembly and are recorded downward (as the well is deepened) rather than upward from the bottom of the hole (as wireline logs are recorded).

Memory Log

This method of data acquisition involves recording the sensor data into a down hole memory, rather than transmitting "Real Time" to surface. There are some advantages and disadvantages to this memory option.

- The tools can be conveyed into wells where the trajectory is deviated or extended beyond the reach of conventional Electric Wireline cables. This can involve a combination of weight to strength ratio of the electric cable over this extended reach. In such cases the memory tools can be conveyed on Pipe or Coil Tubing.
- The type of sensors are limited in comparison to those used on Electric Line, and tend to be focused on the cased hole, production stage of the well. Although there are now developed some memory "Open Hole" compact formation evaluation tool combinations. These tools can be deployed and carried downhole concealed internally in drill pipe to protect them from damage while running in the hole, and then "Pumped" out the end at depth to initate logging. Other basic open hole formation evaluation memory tools are available for use in "Commodity" markets on slickline to reduce costs and operating time.
- In cased hole operation there is normally a "Slick Line" intervention unit. This uses a solid mechanical wire (.82 - .125 inches in OD), to manipulate or otherwise carry out operations in the well bore completion system. Memory operations are often carried out on this Slickline conveyance in preference to mobilizing a full service Electric Wireline unit.
- Since the results are not known until returned to surface, any realtime well dynamic changes cannot be monitored real time. This limits the ability to modify or change the well down hole production conditions accurately during the memory logging by changing the surface production rates. Something that is often done in Eletric Line operations.

- Failure during recording is not known until the memory tools are retrieved. This loss of data can be a major issue on large offshore (expensive) locations. On land locations (e.g. South Texas, US) where there is what is called a "Commodity" Oil service sector, where logging often is without the rig infrastructure. This is less problematic, and logs are often run again without issue.

Information Use

In the oil industry, the well and mud logs are usually transferred in 'real time' to the operating company, which uses these logs to make operational decisions about the well, to correlate formation depths with surrounding wells, and to make interpretations about the quantity and quality of hydrocarbons present. Specialists involved in well log interpretation are called log analysts.

Gamma Ray Logging

Gamma ray logging is a method of measuring naturally occurring gamma radiation to characterize the rock or sediment in a borehole. It is sometimes used in mineral exploration and water-well drilling, but most commonly for formation evaluation in oil and gas well drilling. Different types of rock emit different amounts and different spectra of natural gamma radiation. In particular, shales usually emit more gamma rays than other sedimentary rocks, such as sandstone, gypsum, salt, coal, dolomite, or limestone because radioactive potassium is a common component in their clay content, and because the cation exchange capacity of clay causes them to adsorb uranium and thorium. This difference in radioactivity between shales and sandstones/ carbonate rocks allows the gamma tool to distinguish between shales and non-shales.

The gamma ray log, like other types of well logging, is done simply by lowering an instrument down the hole and recording gamma radiation variation with depth. In the United States, the device most commonly records measurements at 1/2-foot intervals. Gamma radiation is usually recorded in API units, a measurement originated by the petroleum industry. Gamma logs are affected by the diameter of the borehole and the properties of the fluid filling the borehole, but because gamma logs are most often used in a qualitative way, corrections are usually not necessary.

Three elements and their decay chains are responsible for the radiation emitted by rock: potassium, thorium and uranium. Shales often contain potassium as part of their clay content, and tend to absorb uranium and thorium as well. A common gamma-ray log records the total radiation, and cannot distinguish between the radioactive elements, while a spectral gamma ray log can.

For standard GR logs the value measured is calculated from thorium in ppm, Uranium in ppm and potassium in percent. GR API = 8 × Uranium concentration in ppm + 4 × thorium concentration in ppm + potassium concentration in percent.

Due to the weight of uranium concentration in the calculation anomalous concentrations of uranium can cause clean sand reservoirs to appear shaley. Spectral Gamma ray is used to provide an individual reading for each element so anomalies in concentration can be found and interpreted.

An advantage of the gamma log over some other types of well logs is that it works through the steel and cement walls of cased boreholes. Although concrete and steel absorb some of the gamma radiation, enough travels through the steel and cement to allow qualitative determinations.

Sometimes non-shales also have elevated levels of gamma radiation. Sandstone can contain uranium mineralization, potassium feldspar, clay filling, or rock fragments that cause it to have higher-than usual gamma readings. Coal and dolomite may contain absorbed uranium. Evaporite deposits may contain potassium minerals such as carnallite. When this is the case, spectral gamma ray logging can be done to identify these anomalies.

Spectral Logging

Some specialized gamma radiation logging distinguishes the three component decay chains (potassium, uranium, and thorium) by the wavelengths of their characteristic gamma emissions. The characteristic gamma ray line that is associated with each component:

- Potassium : Gamma ray energy 1.46 MeV
- Thorium series: Gamma ray energy 2.62 MeV
- Uranium-Radium series: Gamma ray energy 1.76 MeV.

Another example of the use of spectral gamma ray logs is to identify specific clay types, like Kaolinite or Illite. This can be used for environmental interpretation as Kaolinite forms from Feldspars in tropic soils by leaching of Potassium; and low Potassium readings may thus indicate paleosols. The identification of clay types is also useful for calculating the effective porosity of reservoir rock.

Use in Mineral Exploration

Gamma ray logs are also used in mineral exploration, especially exploration for phosphates, uranium, and potassium salts.

Spontaneous Potential Logging

{{Well loggin The spontaneous potential log, commonly called the self potential log or SP log, is a passive measurement taken by oil industry well loggers to characterise rock formation properties. The log works by measuring small electric potentials (measured in millivolts) between depths in the borehole and a grounded voltage at the surface.

The change in voltage through the well bore is caused by a buildup of charge on the well bore walls. Clays and shales (which are composed predominantly of clays) will generate one charge and permeable formations such as sandstone will generate an opposite one. This build up of charge is, in turn, caused by differences in the salt content of the well bore fluid (drilling mud) and the formation water (connate water). The potential opposite shales is called the baseline, and typically shifts only slowly over the depth of the borehole. Whether the mud contains more or less salt than the connate water will determine the which way the SP curve will deflect opposite a permeable formation. The amplitudes of the line made by the changing SP will vary from formation to formation and will not give a definitive answer to how permeable or the porosity of the formation that it is logging. The SP tool is one of the simplest tools and is generally run as standard when logging a hole, along with the gamma ray. SP data can be used to find:

- Where the permeable formations are
- The boundaries of these formations
- Correlation of formations when compared with data from other analogue wells
- Values for the formation-water resistivity.

The SP curve can be influenced by various factors both in the formation and introduced into the wellbore by the drilling process. These factors can cause the SP curve to be muted or even inverted depending on the situation.

- Formation bed thickness
- Resistivities in the formation bed and the adjacent formations
- Resistivity and make up of the drilling mud
- Wellbore diameter
- The depth of invasion by the drilling mud into the formation.

Mud invasion into the permeable formation can cause the deflections in the SP curve to be rounded off and to reduce the amplitude of thin beds. A smaller wellbore will cause, like a mud filtrate invasion, the deflections on the SP curve to be rounded off and decrease the amplitude opposite thin beds, while a larger diameter wellbore has the opposite effect. If the salinity of the mud filterate is greater than formation water the SP currents will flow in opposite direction. In that case SP deflection will be positive towards to the right positive deflections are observed for fresh water bearing formations.

Resistivity Logging

Resistivity logging is a method of well logging that works by characterizing the rock or sediment in a borehole by measuring its electrical resistivity. Resistivity is a fundamental material property which represents how strongly a material opposes the flow of electric current. In these logs, resistivity is measured using 4 electrical probes to eliminate the resistance of the contact leads. The log must run in holes containing electrically conductive mud or water.

Resistivity logging is sometimes used in mineral exploration and water-well drilling, but most commonly for formation evaluation in oil- and gas-well drilling. Most rock materials are essentially insulators, while their enclosed fluids are conductors. Hydrocarbon fluids are an exception, because they are almost infinitely resistive. When a formation is porous and contains salty water, the overall resistivity will be low. When the formation contains hydrocarbon, or contains very low porosity, its resistivity will be high. High resistivity values may indicate a hydrocarbon bearing formation.

Usually while drilling, drilling fluids invade the formation, changes in the resistivity are measured by the tool in the invaded zone. For this reason, several resistivity tools with different investigation lengths are used to measure the formation resistivity. If water based mud is used and oil is displaced, "deeper" resistivity logs (or those of the "virgin zone") will show lower conductivity than the invaded zone. If oil based mud is used and water is displaced, deeper logs will show higher conductivity than the invaded zone. This provides not only an indication of the fluids present, but also, at least quantitatively, whether the formation is permeable or not.

Use in Mineral Exploration

Resistivity logs are also used in mineral exploration, especially exploration for iron and potassium.

Density Logging

Density logging is a well logging tool determining rock bulk density along a wellbore. This is the overall density of a rock including solid matrix and the fluid enclosed in pores. Geologically, bulk density is a function of the density of the minerals forming a rock (i.e. matrix) and the enclosed volume of free fluids (porosity).

General Tool Design

Modern density logging tools, generally shares the same design principles independent of manufacturer. Tool consist of housing with electronics catridges and a sonde part. Sonde part has a measuring pad, where radiactive source (CS-137 nowadays) is inserted and a number of detectors (Long Space, Short Space, and a "Backscattering" detector).

Measuring pad is connected with a hydraulically operated caliper arm, which is used to measure borehole diameter, and on the same time apply pressure to the pad to ensure it's contact with the borehole. Density measurements are extremely sensitive to shape of the borehole - readings in large washouts are irrelevant due to very shallow depth of investigation of density tools.

Principle

A radioactive source applied to the hole wall emits medium-energy gamma rays into the formation so these gamma rays may be thought of as high velocity particles which collide with the electrons in the formation. At each collision the gamma ray loses some of its energy to the electron, and then continues with diminished energy. This type of interaction is known as Compton scattering. The scattered gamma rays reaching the detector, at the fixed station from the source, are counted as an indication of formation density.

The number of Compton scattering collisions is related directly to the number of the electron density of the formation. Consequently, the electron density determines the response of the density tool. Electron density is related through equation

$$\rho\dot{\epsilon} = \frac{2\rho B \sum Z_{1i}}{M}$$

where $\sum Z_{1i}$ is the atomic numbers of all the atoms making up the molecules in the compound, and M is them molecular weight of the compound.

Inferring Porosity from Bulk Density

Assuming that the measured bulk density (ρ_{bulk}) only depends on matrix density (ρ_{matrix}) and fluid density (ρ_{fluid}), and that these values are known along the wellbore, porosity (φ) can be inferred by the formula

$$\phi = \frac{\rho_{matrix} - \rho_{bulk}}{\rho_{matrix} - \rho_{fluid}}$$

Common values of matrix density ρ_{matrix} (in g/cm^3) are:

- Quartz sand - 2.65
- Limestone - 2.71
- Dolomite - 2.87.

Density of clay minerals is highly variable, and depends on depositional environment, overburden pressure, type of clay mineral and many other factors. It can vary from 2.1 (Montmorillonite) to 2.76 (Chlorite). A fluid bulk density ρ_{fluid} of 1 g/cm^3 is appropriate where the water is fresh; highly saline water has a slightly higher density.

For flushed gas or oil reservoirs, even lower ρ_{fluid} values should be assumed depending on the hydrocarbon density and residual saturation. In some applications hydrocarbons are indicated by the presence of abnormally high log porosities.

Sonic Logging

Sonic logging shows a formation's interval transit time, designated *Dt*. It is a measure of a formation's capacity to transmit sound waves. Geologically, this capacity varies with lithology and rock textures, notably porosity.

Quantitatively, the sonic log is used to evaluate porosity in liquid filled pores. The sonic tool is only capable of measuring travel time. Many relationships between travel time and porosity have been proposed, the most commonly accepted is the Wyllie time average equation.

The equation basically holds that the total travel time recorded on the log is the sum of the time the sonic wave spends travelling the solid part of the rock, called the rock matrix and the time spent travelling through the fluids in the pores.

Use in Mineral Exploration

Sonic logs are also used in mineral exploration, especially exploration for iron, hydrocarbons and potassium.

Caliper Log

A caliper log is a set of measurements of the size and shape of a bore hole commonly made when drilling oil and gas wells. This can be an important indicator of cave ins or shale swelling in the bore hole.

The caliper tool measures the variation in bore hole diameter as it is withdrawn from the bottom of the hole. It is constructed with two or more articulated arms that push against the bore hole wall to take measurements. The arms show variable movements of the cursor by measuring electrical resistance, creating electrical variation. The variation in output is translated into changes of diameter after a simple calibration. The caliper log is printed as a continuous series of values of hole diameter with depth.

Known challenges with caliper logging include borehole spiralling. The position of the drill bit may precess as it drills, leading to spiraling shapes in the wellbore wall, as if the hole had been drilled by a screw. If the arms of the caliper log follow the grooves of the spiral, it will report too high an average diameter. Moving in and out of the grooves, the caliper will give erratic or periodically varying readings.

Mud Logging

Mud logging, also known as hydrocarbon well logging, is the creation of a detailed record (well log) of a borehole by examining the bits of rock or sediment brought to the surface by the circulating drilling medium (most commonly mud). Mud logging is usually performed by a third-party mud logging company. This provides well owners and producers with information about the lithology and fluid content of the borehole while drilling. Historically it is the earliest type of well log. Under some circumstances compressed air is employed as a circulating fluid, rather than mud. Although most commonly used in petroleum exploration, mud logging is also sometimes used when drilling water wells and in other mineral exploration, where drilling fluid is the circulating medium used to lift cuttings out of the hole. In hydrocarbon exploration, hydrocarbon surface gas detectors record the level of natural gas brought up in the mud. A mobile laboratory is situated by the mud logging company near the drilling rig or on deck of an offshore drilling rig, or on a drill ship.

The Service Contracted

A mud logging technician in the modern oil field drilling operation determines positions of hydrocarbons with respect to depth, identifies downhole lithology, monitors natural gas entering the drilling mud stream, and draws well logs for use by oil company geologists. Rock cuttings circulated to the surface in drilling mud are sampled and analyzed.

The mud logging company is normally contracted by the oil company (or operator). They then organize this information in the form of a graphic log, showing the data charted on a representation of the wellbore.

The oil industry representative (or "company man") (also with the tool pusher, and geologist) provides mud loggers their instruction. The mud logging company is contracted specifically as to when to start well-logging activity and what services to provide. Mud logging may begin on the first day of drilling, known as the "spud in" date but is more likely at some later time (and depth) determined by the oil industry geologist's research. The mud logger may also possess logs from wells drilled in the surrounding area. This information (known as "offset data") can provide valuable clues as to the characteristics of the particular geo-strata that the rig crew is about to drill through.

Mud loggers connect various sensors to the drilling apparatus and install specialized equipment to monitor or "log" drill activity. This can be physically and mentally challenging, especially when having to be done during drilling activity. Much of the equipment will require precise calibration or alignment by the mud logger to provide accurate readings. Mud logging technicians observe and interpret the indicators in the mud returns during the drilling process, and at regular intervals log properties such as drilling rate, mud weight, flowline temperature, oil indicators, pump pressure, pump rate, lithology (rock type) of the drilled cuttings, and other data. Mud logging requires a good deal of diligence and attention. Sampling the drilled cuttings must be performed at predetermined intervals, and can be difficult during rapid drilling.

Another important task of the mud logger is to monitor gas levels (and types) and notify other personnel on the rig when gas levels may be reaching dangerous levels, so appropriate steps can be taken to avoid a dangerous well blowout condition. Because of the lag time between drilling and the time required for the mud and cuttings to return to the surface, a modern augmentation has come into use: Measurement while drilling. The MWD technician, often a separate service company employee, logs data in a similar manner but whose data is different in source and content. Most of the data logged by an MWD technician comes from expensive and complex, sometimes electronic, tools that are downhole installed at or near the drill bit.

Scope

Mud logging includes observation and microscopic examination of drill cuttings (formation rock chips), and evaluation of gas hydrocarbon and its constituents, basic chemical and mechanical parameters of drilling fluid or drilling mud (such as chlorides and temperature), as well as compiling other information about the drilling parameters. Then data is plotted on a graphic log called a mud log.

Other real-time drilling parameters that may be compiled include, but are not limited to; rate of penetration (ROP) of the bit (sometimes called the drill rate), pump rate (quantity of fluid being pumped), pump pressure, weight on bit, drill string weight, rotary speed, rotary torque, RPM (Revolutions Per Minute), SPM (Strokes Per Minute) mud volumes, mud weight and mud viscosity. This information is usually obtained by attaching *monitoring devices* to the drilling rig's equipment with a few exceptions such as the mud weight and mud viscosity which are measured by the derrickhand or the mud engineer.

Rate of drilling is affected by the pressure of the column of mud in the borehole and its relative counterbalance to the internal pore pressures of the encountered rock. A rock pressure greater than the mud fluid will tend to cause rock fragments to spall as it is cut and can increase the drilling rate. "D-exponents" are mathematical trend lines which estimate this internal pressure. Thus both visual evidence of spalling and mathematical plotting assist in formulating recommendations for optimum drilling mud densities for both safety (blowout prevention) and economics. (Faster drilling is generally preferred.)

Mud logging is often written as a single word "mudlogging". The finished product can be called a "mud log" or "mudlog". The occupational description is "mud logger" or "mudlogger". In most cases, the two word usage seems to be more common. The mud log provides a reliable *time log* of drilled formations.

Details of the Mud Log

- The *rate of penetration* in is represented by the black line on the left side of the log. The farther to the left that the line goes, the faster the rate of penetration. On this mud log, ROP is measured in feet per hour but on some older, hand drawn mud logs, it is measured in minutes per foot.
- The *porosity* in is represented by the blue line farthest to the left of the log. It indicates the pore space within the rock structure. An analogy would be the holes in a sponge. The oil

and gas resides within this pore space. Notice how far to the left the porosity goes where all the sand (in yellow) is. This indicates that the sand has good porosity. Porosity is not a direct or physical measurement of the pore space but rather an extrapolation from other drilling parameters and therefore not always reliable.

- The *lithology* in is represented by the cyan, gray/black and yellow blocks of colour. Cyan = lime, gray/black = shale and yellow = sand. More yellow represents more sand identified at that depth. The lithology is measured as percentage of the total sample, as visually inspected under a microscope, normally at 10x magnification. These are but a fraction of the different types of formations that might be encountered. (Colour coding is not necessarily standardized among different mud logging companies, though the symbol representation for each are very similar.) In you can see a sample of cuttings under a microscope at 10x magnification after they have been washed off. Some of the larger shale and lime fragments are separated from this sample by running it through sieves and must be considered when estimating percentages. With that in mind this sample would be considered to be about 90% shale, 5% sand and 5% lime (In 5% increments).
- The *gas* in is represented by the green line and is measured in units of ppm (parts per million) as the quantity of total gas, but does not represent the actual quantity of oil or gas the reservoir contains. In the squared-off dash-dot lines just to the right of the sand (in yellow) and left of the gas (in green) represents the heavier hydrocarbons detected. Cyan = C_2 (ethane), purple = C_3 (propane) and blue = C_4 (butane). Detecting and analyzing these heavy gases help to determine the type of oil or gas the formation contains.

Measurement While Drilling

MWD stands for Measurement While Drilling in the oil industry. Is a system developed to perform drilling related measurements downhole and transmit information to the surface while drilling a well. MWD tools are conveyed downhole as part of bottom hole assembly (BHA). The tools are either contained inside a drill collar (sonde type) or are built into the collars themselves.

MWD systems can take several measurements like natural gamma ray, directional survey, tool face, borehole pressure, temperature, vibration, shock, torque etc. Some advanced MWD tools can even measure formation pressure and take formation samples. The MWD also provides the telemetry for operating rotary steering tools (RSTs).

The measured results are stored in MWD tools and some of the results can be transmitted digitally to surface using mud pulser telemetry through the mud or other advanced technology. Certain MWD systems have the capability of receiving encoded control commands which are sent by turning on and off mud pumps and/or changing the rotation speed of drill pipe or by other advanced telemetry technology such as wired pipe.

Types of Information Transmitted

Directional Information

MWD tools are generally capable of taking directional surveys in real time. The tool uses accelerometers and magnetometers to measure the inclination and azimuth of the wellbore at that location, and they then transmit that information to the surface. With a series of surveys at appropriate intervals (anywhere from every 30 ft (ie 10m) to every 500 ft), the location of the wellbore can be calculated.

By itself, this information allows operators to prove that their well does not cross into areas that they are not authorized to drill. However, due to the cost of MWD systems, they are not generally used on wells intended to be vertical. Instead, the wells are surveyed after drilling through the use of Multishot Surveying Tools lowered into the drillstring on slickline or wireline. The primary use of real-time surveys is in Directional Drilling. For the Directional Driller to steer the well towards a target zone, he must know where the well is going, and what the effects of his steering efforts are. MWD tools also generally provide toolface measurements to aid in directional drilling using downhole mud motors with bent subs or bent housings.

Drilling Mechanics Information

MWD tools can also provide information about the conditions at the drill bit. This may include:

- Rotational speed of the drillstring
- Smoothness of that rotation
- Type and severity of any vibration downhole
- Downhole temperature
- Torque and Weight on Bit, measured near the drill bit
- Mud flow volume.

Use of these information can allow the operator to drill the well more efficiently, and to ensure that the MWD tool and any other downhole tools, such as Mud Motors, Rotary Steerable Systems, and LWD tools, are operated within their technical specifications to prevent tool failure. This information also is valuable to Geologists responsible for the well information about the formation which is being drilled.

Formation Properties

Many MWD tools, either on their own, or in conjunction with separate Logging While Drilling tools, can take measurements of formation properties. At the surface, these measurements are assembled into a log, similar to one obtained by wireline logging.

LWD Logging While Drilling tools are able to measure a suite of geological characteristics including- density, porosity, resistivity, acoustic-caliper, inclination at the drill bit (NBI), magnetic resonance and formation pressure.

The MWD tool allows these measurements to be taken and evaluated while the well is being drilled. This makes it possible to perform Geosteering, or Directional Drilling based on measured formation properties, rather than simply drilling into a preset target.

Most MWD tools contain an internal Gamma Ray sensor to measure natural Gamma Ray values. This is because these sensors are compact, inexpensive, reliable, and can take measurements through unmodified drill collars. Other measurements often require separate Logging While Drilling tools, which communicate with the MWD tools downhole through internal wires. Measurement while drilling can be cost-effective in exploration wells, particularly in areas of the Gulf of Mexico where wells are drilled in areas of salt diapirs. The resistivity log will detect penetration into salt, and early detection prevents salt damage to bentonite drilling mud.

Data Transmission Methods

Mud Pulse Telemetry

This is the most common method of data transmission used by MWD (Measurement While Drilling) tools. Downhole a valve is operated to restrict the flow of the drilling mud (slurry) according to the digital information to be transmitted. This creates pressure fluctuations representing the information. The pressure fluctuations propagate within the drilling fluid towards the surface where they are received from pressure sensors. On the surface, the received pressure signals are processed by computers to reconstruct the information. The technology is available in three varieties - *positive* pulse, *negative* pulse, and *continuous wave.*

Positive Pulse

Positive Pulse tools briefly close and open the valve to restrict the mud flow within the drill pipe. This produces an increase in pressure that can be seen at surface. Line codes are used to represent the digital information in form of pulses.

Negative Pulse

Negative pulse tools briefly open and close the valve to release mud from inside the drillpipe out to the annulus. This produces a decrease in pressure that can be seen at surface. Line codes are used to represent the digital information in form of pulses.

Continuous Wave

Continuous wave tools gradually close and open the valve to generate sinusoidal pressure fluctuations within the drilling fluid. Any digital modulation scheme with a continuous phase can be used to impose the information on a carrier signal. The most widely used modulation scheme is continuous phase modulation. When underbalanced drilling is used, mud pulse telemetry can become unusable. This is because usually in order to reduce the equivalent density of the drilling mud a compressible gas is injected into the mud. This causes high signal attenuation which drastically reduces the ability of the mud to transmit pulsed data. In this case it is necessary to use methods different from mud pulse telemetry, such as electromagnetic waves propagating through the formation or wired drill pipe telemetry.

Current mud pulse telemetry technology offers a bandwidths of up to 40 bps. The data rate drops with increasing length of the wellbore and is typically as low as 1.5 bps - 3.0 bps. (bits per second) at a depth of 35,000 ft - 40,000 ft (10668 m - 12192 m).

Surface to down hole communication is typically done via changes to drilling parameters, i.e. change of the rotation speed of the drill string or change of the mud flow rate. Making changes to the drilling parameters in order to send information can require interruption of the drilling process, which is unfavorable due to the fact that it causes non-productive time.

Electromagnetic Telemetry (EM Tool)

These tools incorporate an electrical insulator in the drillstring. To transmit data the tool generates an altered voltage difference between the top part (the main drillstring, above the insulator), and the bottom part (the drill bit, and other tools located below the insulator of the MWD tool).

On surface a wire is attached to the wellhead, which makes contact with the drillpipe at the surface. A second wire is attached to a rod driven into the ground some distance away. The wellhead and the ground rod form the two electrodes of a dipole antenna. The voltage difference between the two electrodes is the receive signal that is decoded by a computer.

The EM tool generates voltage differences between the drillstring sections in the pattern of very low frequency (2–12 Hz) waves. The data is imposed on the waves through digital modulation.

This system generally offers data rates of up to 10 bits per second. In addition, many of these tools are also capable of receiving data from the surface in the same way, while mud pulse-based tools rely on changes in the drilling parameters, such as rotation speed of the drillstring or the mud flow rate, to send information from the surface to downhole tools. Making changes to the drilling parameters in order to send information to the tools generally interrupts the drilling process, causing lost time.

Compared to mud pulse telemetry, electronic pulse telemetry is more effective in certain specialized situations, such as underbalanced drilling or when using air as drilling fluid. However, it generally falls short when drilling exceptionally deep wells, and the signal can lose strength rapidly in certain types of formations, becoming undetectable at only a few thousand feet of depth.

Wired Drill Pipe

Several oilfield service companies are currently developing wired drill pipe systems. These systems use electrical wires built into every component of the drillstring, which carry electrical signals directly to the surface. These systems promise data transmission rates orders of magnitude greater than anything possible with mud pulse or electromagnetic telemetry, both from the downhole tool to the surface, and from the surface to the downhole tool.

The IntelliServ wired pipe network, offering data rates upwards of 1 megabit per second, became commercial in 2006. Representatives from BP America, StatoilHydro, Baker Hughes INTEQ, and Schlumberger presented three success stories using this system, both onshore and offshore, at the March, 2008 SPE/IADC Drilling Conference in Orlando, Florida.

Retrievable Tools

MWD tools may be semi-permanently mounted in a drill collar (only removable at servicing facilities), or they may be self-contained and wireline retrievable.

Retrievable tools, sometimes known as *Slim Tools*, can be retrieved and replaced using wireline through the drill string. This generally allows the tool to be replaced much faster in case of failure, and it allows the tool to be recovered if the drillstring becomes stuck. Retrievable tools must be much smaller, usually about 2 inches or less in diameter, though their length may be 20 feet or more. The small size is necessary for the tool to fit through the drillstring, however, it also limits the tool's capabilities. For example, slim tools are not capable of sending data at the same rates as collar mounted tools, and they are also more limited in their ability to communicate with and supply electrical power to other LWD tools.

Collar-mounted tools, also known as *Fat Tools*, cannot generally be removed from their drill collar at the wellsite. If the tool fails, the entire drillstring must be pulled out of the hole to replace it. However, without the need to fit through the drillstring, the tool can be larger and more capable.

The ability to retrieve the tool via wireline is often useful. For example, if the drillstring becomes stuck in the hole, then retrieving the tool via wireline will save a substantial amount of money compared to leaving it in the hole with the stuck portion of the drillstring. However, there are some limitations on the process.

Limitations

Retrieving a tool using wireline is not necessarily faster than pulling the tool out of the hole. For example, if the tool fails at 1,500 ft (460 m) while drilling with a triple rig (able to trip 3 joints of pipe, or about 90 ft (30 m) feet, at a time), then it would generally be faster to pull the tool out of the hole than it would be to rig up wireline and retrieve the tool, especially if the wireline unit must be transported to the rig.

Wireline retrievals also introduce additional risk. If the tool becomes detached from the wireline, then it will fall back down the drillstring. This will generally cause severe damage to the tool and the drillstring components in which it seats, and will require the drillstring to be pulled out of the hole to replace the failed components, thus resulting in a greater total cost than pulling out of the hole in the first place. The wireline gear might also fail to latch onto the tool, or in the case of a severe failure, might bring only a portion of the tool to the surface. This would require the drillstring to be pulled out of the hole to replace the failed components, thus making the wireline operation a waste of time.

Petroleum Engineering

Petroleum engineering is an engineering discipline concerned with the activities related to the production of hydrocarbons, which can be either crude oil or natural gas. Subsurface activities are deemed to fall within the *upstream* sector of the oil and gas industry, which are the activities of finding and producing hydrocarbons. Refining and distribution to a market are referred to as the *downstream* sector. Exploration, by earth scientists, and petroleum engineering are the oil and gas industry's two main subsurface disciplines, which focus on maximizing economic recovery of hydrocarbons from subsurface reservoirs.

Petroleum geology and geophysics focus on provision of a static description of the hydrocarbon reservoir rock, while petroleum engineering focuses on estimation of the recoverable volume of this resource using a detailed understanding of the physical behaviour of oil, water and gas within porous rock at very high pressure.

The combined efforts of geologists and petroleum engineers throughout the life of a hydrocarbon accumulation determine the way in which a reservoir is developed and depleted, and usually they have the highest impact on field economics. Petroleum engineering requires a good knowledge of many other related disciplines, such as geophysics, petroleum geology, formation evaluation (well logging), drilling, economics, reservoir simulation, well engineering, artificial lift systems, and oil and gas facilities engineering.

Petroleum engineering has historically been one of the highest paid engineering disciplines; although there is a tendency for mass layoffs when oil prices decline. In a June 4th, 2007 article, Forbes.com reported that petroleum engineering was the 24th best paying job in the United States. The 2010 National Association of Colleges and Employers survey showed petroleum engineers as the highest paid 2010 graduates at an average $125,220 annual salary. For individuals with experience, salaries can go from $170,000 to $260,000 annually.

Petroleum engineering has become a technical profession that involves extracting oil in increasingly difficult situations as much of the "low hanging fruit" of the world's oil fields has been found and depleted. Improvements in computer modelling, materials and the application of statistics, probability analysis, and new technologies like horizontal drilling and enhanced oil recovery, have drastically improved the toolbox of the petroleum engineer in recent decades. Deep-water, arctic and desert conditions are commonly contended with. High Temperature and High Pressure (HTHP) environments have become increasingly commonplace in operations and require the petroleum engineer to be savvy in topics as wide ranging as thermo-hydraulics, geomechanics, and intelligent systems.

The Society of Petroleum Engineers (SPE) is the largest professional society for petroleum engineers and publishes much information concerning the industry. Petroleum engineering education is available at 17 universities in the United States and many more throughout the world - primarily in oil producing regions - and some oil companies have considerable in-house petroleum engineering training classes.

Types

Petroleum engineers divide themselves into several types:

- Reservoir engineers work to optimize production of oil and gas via proper well placement, production rates, and enhanced oil recovery techniques.
- Drilling engineers manage the technical aspects of drilling exploratory, production and injection wells.
- Production engineers, including subsurface engineers, manage the interface between the reservoir and the well, including perforations, sand control, downhole flow control, and downhole monitoring equipment; evaluate artificial lift methods; and also select surface equipment that separates the produced fluids (oil, natural gas, and water).

Reservoir Engineering

Reservoir engineering is a branch of petroleum engineering that applies scientific principles to the drainage problems arising during the development and production of oil and gas reservoirs so as to obtain a high economic recovery. The working tools of the reservoir engineer are subsurface geology, applied mathematics, and the basic laws of physics and chemistry governing the behaviour of liquid and vapour phases of crude oil, natural gas, and water in reservoir rock.

Of particular interest to reservoir engineers is generating accurate reserves estimates for use in financial reporting to the SEC and other regulatory bodies. Other job responsibilities include numerical reservoir modelling, production forecasting, well testing, well drilling and workover planning, economic modelling, and PVT analysis of reservoir fluids.

Reservoir engineers also play a central role in field development planning, recommending appropriate and cost effective reservoir depletion schemes such as waterflooding or gas injection to maximize hydrocarbon recovery. Due to legislative changes in many hydrocarbon producing countries, they are also involved in the design and implementation of carbon sequestration projects in order to minimize the emission of greenhouse gases.

Types

Reservoir engineers often specialize in two areas:

- Surveillance (or production) engineering, i.e. monitoring of existing fields and optimization of production and injection rates. Surveillance engineers typically use analytical and empirical techniques to perform their work, including decline curve analysis, material balance modelling, and inflow/outflow analysis.
- Simulation modelling, i.e. the conduct of reservoir simulation studies to determine optimal development plans for oil and gas reservoirs.

Chapter 4

Exploration Geophysics

Exploration geophysics is the applied branch of geophysics which uses surface methods to measure the physical properties of the subsurface Earth, in order to detect or infer the presence and position of ore minerals, hydrocarbons, geothermal reservoirs, groundwater reservoirs, and other geological structures.

Exploration geophysics is the practical application of physical methods (such as seismic, gravitational, magnetic, electrical and electromagnetic) to measure the physical properties of rocks, and in particular, to detect the measurable physical differences between rocks that contain ore deposits or hydrocarbons and those without.

Exploration geophysics can be used to directly detect the target style of mineralisation, via measuring its physical properties directly. For example one may measure the density contrasts between iron ore and silicate wall rocks, or may measure the electrical conductivity contrast between conductive sulfide minerals and barren silicate minerals.

Geophysical Methods

The main techniques used are:

1. Seismic methods, such as reflection seismology, seismic refraction, and seismic tomography.
2. Geodesy and gravity techniques, including gravity gradiometry.
3. Magnetic techniques, including aeromagnetic surveys.
4. Electrical techniques, including electrical resistivity tomography and induced polarization.

5. Electromagnetic methods, such as magnetotellurics, ground penetrating radar and transient/time-domain electromagnetics.
6. Borehole geophysics, also called well logging.
7. Remote sensing techniques, including hyperspectral.

Many other techniques, or methods of integration of the above techniques, have been developed and are currently used. However these are not as common due to cost-effectiveness, wide applicability and/or uncertainty in the results produced.

Uses

Exploration geophysics is also used to map the subsurface structure of a region, to elucidate the underlying structures, spatial distribution of rock units, and to detect structures such as faults, folds and intrusive rocks. This is an indirect method for assessing the likelihood of ore deposits or hydrocarbon accumulations.

Methods devised for finding mineral or hydrocarbon deposits can also be used in other areas such as monitoring environmental impact, imaging subsurface archaeological sites, ground water investigations, subsurface salinity mapping, civil engineering site investigations and interplanetary imaging.

Mineral Exploration

Magnetometric surveys can be useful in defining magnetic anomalies which represent ore (direct detection), or in some cases gangue minerals associated with ore deposits (indirect or inferential detection).

The most direct method of detection of ore via magnetism involves detecting iron ore mineralisation via mapping magnetic anomalies associated with banded iron formations which usually contain magnetite in some proportion.

Skarn mineralisation, which often contains magnetite, can also be detected though the ore minerals themselves would be non-magnetic. Similarly, magnetite, hematite and often pyrrhotite are common minerals associated with hydrothermal alteration, and this alteration can be detected to provide an inference that some mineralising hydrothermal event has affected the rocks.

Gravity surveying can be used to detect dense bodies of rocks within host formations of less dense wall rocks. This can be used to directly detect Mississippi Valley Type ore deposits, IOCG ore deposits, iron ore deposits, skarn deposits and salt diapirs which can form oil and gas traps.

Electromagnetic (EM) surveys can be used to detect a wide variety of base metal sulphide deposits via detection of conductivity anomalies which can be generated around sulphide bodies in the subsurface. EM surveys can also be used to detect palaeochannel-hosted uranium deposits which are associated with shallow aquifers, which often respond to EM surveys in conductive overburden. This is an indirect inferential method of detecting mineralisation.

Regional EM surveys are conducted via airborne methods, using either fixed-wing aircraft or helicopter-borne EM rigs. Surface EM methods are based mostly on Transient EM methods using surface loops with a surface receiver, or a downhole tool lowered into a borehole which transects a body of mineralisation. These methods can map out sulphide bodies within the earth in 3 dimensions, and provide information to geologists to direct further exploratory drilling on known mineralisation. Surface loop surveys are rarely used for regional exploration, however in some cases such surveys can be used with success (e.g.; SQUID surveys for nickel ore bodies).

Electric-resistance methods such as induced polarization methods can be useful for directly detecting sulfide bodies, coal and resistive rocks such as salt and carbonates.

Oil and Gas

Seismic reflection techniques are the most widely used geophysical technique in hydrocarbon exploration. They are used to map the subsurface distribution of stratigraphy and its structure which can be used to delineate potential hydrocarbon accumulations. Well logging is another widely used technique as it provides necessary high resolution information about rock and fluid properties in a vertical section, although they are limited in areal extent. This limitation in areal extent is the reason why seismic reflection techniques are so popular; they provide a method for interpolating and extrapolating well log information over a much larger area.

Gravity and magnetics are also used, with considerable frequency, in oil and gas exploration. These can be used to determine the geometry and depth of covered geological structures including uplifts, subsiding basins, faults, folds, igneous intrusions and salt diapirs due to their unique density and magnetic susceptibility signatures compared to the surrounding rocks. Remote sensing techniques, specifically hyperspectral imaging, have been used to detect hydrocarbon microseepages using the spectral signature of geochemically altered soils and vegetation.

Magnetotellurics and Controlled source electro-magnetics can provide pseudo-direct detection of hydrocarbons by detecting resistivity changes. It can also complement seismic data when imaging below salt.

Civil Engineering

Ground penetrating radar is used within civil construction and engineering for a variety of uses, including detection of utilities (buried water, gas, sewerage, electrical and telecommunication cables), mapping of soft soils and overburden for geotechnical characterization, and other similar uses. Civil engineering can also use remote sensing information for topographical mapping, planning and environmental impact assessment. Airborne electromagnetic surveys are also used to characterize soft sediments in planning and engineering roads, dams and other structures.

Magnetotellurics has proven useful for delineating groundwater reservoirs, mapping faults around areas where hazardous substances are stored (e.g. nuclear power stations and nuclear waste storage facilities), and earthquake precursor monitoring in areas with major structures such as hydro-electric dams subject to high levels of seismic activity.

Archaeology

Ground penetrating radar can be used to map buried artifacts, such as graves, mortuaries, wreck sites, and other shallowly buried archaeological sites. Ground magnetometric surveys can be used for detecting buried ferrous metals, useful in surveying shipwrecks, modern battlefields strewn with metal debris, and even subtle disturbances such as large-scale ancient ruins.

Sonar systems can be used to detect shipwrecks.

Forensics

Ground penetrating radar can be used to detect grave sites.

Unexploded Ordnance Detection

Magnetic and electromagnetic surveys can be used to locate unexploded ordnance.

Prospective Resources

Arctic Prospective Resources

A 2008 United States Geological Survey estimates that areas north of the Arctic Circle have 90 billion barrels (1.4×10^{10} m^3) of undiscovered, technically recoverable oil and 44 billion barrels (7.0×10^9 m^3) of natural gas liquids in 25 geologically defined areas thought to have potential for petroleum. This represented 13% of the expected undiscovered oil in the world. Of the estimated totals, more than half of the undiscovered oil resources were estimated to occur in just three geologic provinces—Arctic Alaska, the Amerasia Basin, and the East Greenland Rift Basins.

More than 70% of the mean undiscovered oil resources was estimated to occur in five provinces: Arctic Alaska, Amerasia Basin, East Greenland Rift Basins, East Barents Basins, and West Greenland–East Canada. It was further estimated that approximately 84% of the oil and gas would occur offshore. The USGS did not consider economic factors such as the effects of permanent sea ice or oceanic water depth in its assessment of undiscovered oil and gas resources. This assessment was lower than a 2000 survey, which had included lands south of the Arctic Circle.

Miscellaneous Prospective Resources

In October 2009, the USGS updated the Orinoco tar sands (Venezuela) value to 513 billion barrels (8.16×10^{10} m^3).

The Exclusive Economic Zone is a delineated offshore area, mostly to the west and north of Cuba, which, under international agreements, is owned by Cuba. This 112,000-square-kilometer zone has been divided into 59 exploration blocks. A 2004 joint partnership between a Spanish oil company and Cuba's state oil company (CUPET) estimated Cuba's off-shore reserves to be able to produce ultimately between 4.6 and 9.3 billion barrels of crude oil.

The US Geological Survey (USGS) estimates that Cuba has resources up to 9 billion barrels (1.4×10^9 m^3) of oil. In October 2008, the Cuban government announced that it had discovered oil basins that would double its total oil resources to 20 billion barrels (3.2×10^9 m^3). Note that, consistent with the definitions of reserves and resources as given in the above sections, at this stage (June 2009), given no commercial discoveries to date, all recoverable oil estimates in the Exclusive Economic Zone (EEZ) are prospective resources estimates and not reserves estimates.

Passive Seismic

Passive seismic is the detection of natural low frequency earth movements, usually with the purpose of discerning geological structure and locate underground oil, gas, or other resources. Usually the data listening is done in multiple measurement points that are separated by several hundred meters, over periods of several hours to several days, using portable seismometers. The conclusions about the geological structure are based on the spectral analysis or on the mathematical reconstruction of the propagation and possible sources of the observed seismic waves. If the latter is planned, data are usually acquired in multiple (in the ideal case - all) points simultaneously, using so called synchronized lines. Reliability of the time reverse modelling can be further increased using results of reflection seismology about the distribution of the sound speed in the underground volume.

Passive seismic usually focuses on a low frequency signals (0 to 10 Hz) and is sometimes called the "low frequency" seismology. The seismometers record movements in all 3 possible directions independently (such devices also have other application areas like long-term measurement stations). When needed, data acquisition is also done under water, using waterproof devices that measure earth movements at the bottom of the sea. Geophones are almost never used due to their limited sensitivity.

The survey using this method is very different from the conventional survey that is usually based on reflection seismology. The conventional survey consists of numerous measurements that are spatially very close together and relatively short (frequently minutes only). The passive seismic survey has much less measurements but they are frequently recorded for days. Local time must be taken into consideration, picking intervals with less human induced noise. Even relatively distant earthquakes are visible in the recorded spectrograms and must also be excluded from analysis

The similar method have been also applied in another planets. For instance, during Apollo missions, the Passive Seismic Experiment sensors were deployed that detected lunar "moonquakes" and provided information about the internal structure of the Moon.

Passive seismic is much cheaper than well drilling. It is also more cheaper and environment friendly than active seismic which requires the strong source of the seismic waves (like underground explosion) to predict the structure. In some cases it may be the only method for which the land access is granted by the land owner. However despite the method has been successfully applied in many parts of the world, this approach is currently less reliable as the scientific methods are still largerly under development.

Magnetometer

A magnetometer is a measuring instrument used to measure the strength or direction of a magnetic field either produced in the laboratory or existing in nature. Some countries such as the USA, Canada and Australia classify the more sensitive magnetometers as military technology, and control their distribution.

The International System of Units unit of measure for the strength of a magnetic field is the tesla. As this is a very large unit, workers in the earth sciences commonly use the nanotesla (nT) as their working unit of measure. Engineers often measure magnetic fields in Gauss. 1 Gauss = 100,000 nT or 1 Gauss = 100,000 gamma.

The Earth's magnetic field (the magnetosphere) is a potential field. It varies both temporally and spatially for various reasons, including inhomogeneity of rocks and interaction between charged particles from the Sun and the magnetosphere.

The earth's magnetic field is relatively weak. A simple magnet that may be purchased in a hardware store produces a field many hundreds of times stronger than the earth's field. The earth's magnetic field varies from around 20,000 nT near the equator to 80,000 nT near the poles. It also varies with time. There is a daily variation of around 30 nT at mid latitudes and hundreds of nT at the poles. Geomagnetic storms can cause much larger variations.

Magnetometers, which measure magnetic fields, are distinct from metal detectors, which detect hidden metals by their conductivity. When used for detecting metals, a magnetometer can detect only magnetic (ferrous) metals, but can detect such metals buried much deeper than a metal detector. Magnetometers are capable of detecting large objects like cars at tens of meters, while a metal detector's range is unlikely to exceed 2 meters.

Uses

Magnetometers have a very diverse range of applications from locating submarines and Spanish Galleons, positioning weapons systems, detecting unexploded ordenance, locating toxic waste drums, heart beat monitors, sensors in anti-locking brakes, weather prediction (via solar cycles), depths of steel pylons, drill guidance systems, locating hazards for tunnel boring machines, archaeology, Plate Tectonics, finding a wide range of mineral deposits and geological structures, hazards in coal mines, to radio wave propagation and planetary exploration. And there are many more applications.

Depending on the application, magnetometers can be deployed in spacecraft, aeroplanes (fixed wing), helicopters (stinger and bird), on the ground (backpack), towed at a distance behind quad bikes (sled or trailer), lowered into boreholes (tool, probe or sonde) and towed behind boats (tow fish). Magnetometers applied to the study the earth are called geophysical surveys — a term that also embraces a wide range of other geophysical techniques including gravity, seismic refraction, seismic reflection, electromagnetics (EM), Induced Polarisation (IP), Magneto-Tellurics (MT), Controlled Source Magneto-Tellurics (CSAMT), Sub-audio Magnetics (SAM), Mise-a-la-Masse, Resistivity, Self Potential (SP) and Very Low Frequency (VLF).

Archaeology

Magnetometers are also used to detect archaeological sites, shipwrecks and other buried or submerged objects. Fluxgate gradiometers are popular due to their compact configuration and relatively low cost. Gradiometers enhance shallow features and negate the need for a base station. Caesium and Overhauser magnetometers are also very effective when used as gradiometers or as single-sensor systems with base stations.

The TV program 'Time Team' popularised 'geophys' including magnetics for archaeological work. Targets include fire hearths, walls of baked bricks, magnetic stones (basalts, granites). Walking tracks and roadways can sometimes be mapped with differential compaction in magnetic soils and/or disturbances in clays such as on the Great Hungarian Plain. Ploughed fields behave as sources of magnetic noise in such surveys.

Auroras

Magnetometers can give an indication of possible auroral activity before one can see the light from the aurora. A grid of magnetometers around the world constantly measures the effect of the solar wind on the Earth's magnetic field, which is published on the K-index.

Coal Exploration

Whilst magnetometers can be used to help map basin shape at a regional scale, they are more commonly used to map hazards to coal mining including basaltic intrusions (dykes, sills and volcanic plugs) that destroy resources and wreak havoc with longwall mining equipment. Magnetometers can also locate faults and burn zones (ignited by lightning). and map siderite – an impurity in some coal.

The best survey results are achieved on the ground in high-resolution surveys (10 m line spacing, 0.5 m station spacing). Borehole magnetometers such as the Ferret can also assist when coal seams are deep; with multiple sills and/or looking beneath surface basalt flows.

Modern surveys generally use magnetometers with GPS to record the magnetic field and locations automatically. The data are corrected using data from a second magnetometer that is left stationary during the survey. This second magnetometer (called the base station) records the change in the earths magnetic field during the time of the survey.

Directional Drilling

They are used in directional drilling for oil or gas to detect the azimuth of the drilling tools near the drill bit. They are most often paired up with accelerometers in drilling tools so that both the inclination and azimuth of the drill bit can be found.

Military

Because a magnetometer can be used to detect a submarine, magnetometers are a classified technology in countries such as Australia, Canada, and the USA.

For defensive purposes, navies use arrays of magnetometers laid across sea floors in strategic locations (i.e. around ports) to monitor submarine activity. The Russian 'Goldfish' (titanium submarines) were designed and built at great expense to thwart such systems (pure titanium is non-magnetic).

Military submarines are degaussed by passing through large underwater loops at regular intervals in a bid to escape detection by sea-floor monitoring systems, magnetic anomaly detectors, and mines that trigger on magnetic anomalies. Submarines are never completely de-magnetised. It is possible to tell how deep a submarine has been diving by measuring its magnetic field, because the pressure distorts the steel and changes the field. Heating can also change the magnetization of steel.

Submarines tow long sonar arrays to listen for ships — they can even recognise different propeller noises. The sonar arrays need to be accurately positioned so they can triangulate direction to targets (e.g. ships). The arrays do not tow in a straight line, so fluxgate magnetometers are used to orient each sonar node in the array.

Fluxgates can also be used in weapons navigation systems, but have been largely superseded by GPS and ring laser gyroscopes.

Magnetometers such as the German Forster are used to locate ferrous ordnance. Cesium and Overhauser magnetometers are used to locate and help clean up old bombing/test ranges.

UAV payloads also include magnetometers for a range of defensive and aggressive tasks.

Mineral Exploration

Mineral exploration is one of the major commercial drivers and users of magnetometers. Magnetometers are one of the prime tools used to locate world class deposits of gold, silver, copper, iron, tin, platinum and diamonds.

Quarry/Gemstone applications include mapping 'Blue Metal' for concrete aggregate and roadbase as well as sapphires, rubies and opal bearing structures.

First world countries such as Australia, Canada and USA invest heavily in systematic airborne magnetic surveys of their respective continents (and surrounding oceans) to help map geology and leverage the discovery of mineral deposits. They use airplanes such as the Shrike Commander.

Such aeromag surveys are typically undertaken on 400 m line spacing at 100 m elevation with readings every 10 meters or more. To overcome the asymmetry in the data density, data is interpolated between lines (usually 5 times) and data along the line is averaged. Such data would be gridded to a 80 m x 80 m pixel size then image processed using a program like ERMapper. At an exploration lease scale, the survey may be followed by a more detailed helimag or crop duster style fixed wing at 50 m line spacing and 50 m elevation (terrain permitting) – the image would be gridded on a 10 x 10 m pixel offering 64 times the resolution.

Where targets are shallow (<200 m), aeromag anomalies may be followed up with ground magnetic surveys on 10 m to 50 m line spacing with 1 m station spacing to give the best detail (2 m to 10 pixel grid) or 25 times the resolution prior to drilling.

Magnetic fields from magnetic orebodies fall off with the inverse distance cubed (dipole target) or at best inverse distance squared (magnetic monopole target). One analogy to the resolution-with-distance is a car driving at night with lights on. At 400 m one sees one glowing haze — as one gets closer one sees two headlights then the left blinker.

There are many challenges interpreting magnetic data for mineral exploration. Multiple targets mix together like multiple heat sources. Unlike light, there is no magnetic telescope to focus fields. We measure the combination of multiple sources at the surface. We also do not know the geometry, depth or magnetisation direction (remanence) of the targets. We can produce multiple models the explain the data — the classic ambiguity problem.

Potent by Geophysical Software Solutions is a leading magnetic (and gravity) interpretation package used extensively in the Australian exploration industry.

Magnetometers assist mineral explorers both directly (i.e. gold mineralisation associated with magnetite, diamonds in kimberlite pipes) and more commonly by indirect means such as mapping geological structures conducive to mineralisation (i.e. shear zones and alteration haloes around granites).

Mobile Telephones

Many smartphones contain magnetometers. There are compass apps which show direction.:

Researchers at Deutsche Telekom have used magnetometers embedded in mobile devices to permit touchless 3-D interaction. Their interaction framework, called MagiTact, tracks changes to the magnetic field around a cellphone to identify different gestures made by a hand holding or wearing a magnet.

Oil Exploration

Seismic methods are preferred to magnetometers for oil exploration. Aeromag surveys can be used for basin shape, and locating faults.

Oil deposits can leak hydrocarbons which find their way up fractures in the ground to be eaten by bugs at or near the surface. The bugs can precipitate magnetite from haematite producing subtle magnetic anomalies. Such anomalies are best mapped by ground based magnetometers.

Spacecraft

A three-axis fluxgate magnetometer was part of the Mariner 2 and Mariner 10 missions. A dual technique magnetometer is part of the Cassini–Huygens mission to explore Saturn. This system is composed of a vector helium and fluxgate magnetometers. Magnetometers are also a component instrument on the Mercury MESSENGER mission. A magnetometer can also be used by satellites like GOES to measure both the magnitude and direction of a planet's or moon's magnetic field.

Types

Magnetometers can be divided into two basic types:

- Scalar magnetometers measure the total strength of the magnetic field to which they are subjected, and

- Vector magnetometers have the capability to measure the component of the magnetic field in a particular direction, relative to the spatial orientation of the device.

Magnetometers can also be classified as "AC" types that measure fields that vary relatively rapidly in time, and "DC" types that measure fields that vary only slowly, if at all (quasi-static). AC magnetometers find use in electromagnetic systems (such as magnetotellurics), and DC magnetometers are used for detecting mineralization and corresponding geological structures.

Vector Magnetometers

A vector is a mathematical entity with both magnitude and direction. The earth's magnetic field at a given point is a vector; it is not just a numerical value, but also points in a specific direction. The direction is three-dimensional, not just north-south but also an inclination from the horizontal. A magnetic compass is designed to give a horizontal bearing direction; a *vector magnetometer* measures the magnitude and direction of the total magnetic field. An example of such a device is a Variometer used in magnetic observatories for monitoring the ionosphere.

Three orthogonal sensors are required to measure the components of the magnetic field in all three dimensions. Vector magnetometers electronically measure one or more components of the magnetic field. Using three orthogonal magnetometers, both azimuth and dip (inclination) can be measured. By taking the square root of the sum of the squares of the components the total magnetic field strength (also called total magnetic intensity, TMI) can be calculated by Pythagoras's theorem.

Examples of vector magnetometers are fluxgates, superconducting quantum interference devices (SQUIDs), and the atomic SERF magnetometer. Fluxgates come in the following 'flavors': ring core, ractrack, rod and Vacquier depending on the geometry of the ferrite cores.

They are subject to temperature drift and the dimensional instability of the ferrite cores. They also require leveling to obtain component information, unlike total field (scalar) instruments. For these reasons they are no longer used for mineral exploration.

Scalar Magnetometers

Scalar magnetometers measure the total magnetic field strength but not its direction. These include Proton Precession, Overhauser, and a range of Alkali vapour instruments including Cesium, Helium and Potassium.

A magnetograph is a special magnetometer that continuously records data.

Rotating Coil Magnetometer

The magnetic field induces a sine wave in a rotating coil. The amplitude of the signal is proportional to the strength of the field, provided it is uniform, and to the sine of the angle between the rotation axis of the coil and the field lines. This type of magnetometer is obsolete.

Hall Effect Magnetometer

The most common magnetic sensing devices are solid-state Hall effect sensors. These sensors produce a voltage proportional to the applied magnetic field and also sense polarity.

They are used in applications where the magnetic field strength is relatively large — for example in Anti-lock braking system in cars to sense wheel rotation speed via slots in the wheel disks.

Proton Precession Magnetometer

Proton precession magnetometers, also known as *proton magnetometers*, PPM's or simply mags, measure the resonance frequency of protons (hydrogen nuclei) in the magnetic field to be measured, due to nuclear magnetic resonance (NMR). Because the precession frequency depends only on atomic constants and the strength of the ambient magnetic field, the accuracy of this type of magnetometer is very good.

A direct current flowing in a solenoid creates a strong magnetic field around a hydrogen-rich fluid (kerosine, and decane is popular — even water can be used), causing some of the protons to align themselves with that field. The current is then interrupted, and as protons realign themselves with ambient magnetic field, they precess at a frequency that is directly proportional to the magnetic field. This produces a weak rotating magnetic field that is picked up by a (sometimes separate) inductor, amplified electronically, and fed to a digital frequency counter whose output is typically scaled and displayed directly as field strength or output as digital data.

The relationship between the frequency of the induced current and the strength of the magnetic field is called the *proton gyromagnetic ratio*, and is equal to 0.042576 Hz/nT. The accuracy of PPM's is this limited by the accuracy of this constant.

The frequency of Earth's field NMR (EFNMR) for protons varies between approximately 900 Hz near the equator to 4.2 kHz near the geomagnetic poles. These magnetometers can be moderately sensitive if several tens of watts are available to power the aligning process. Measuring once per second, standard deviations in the readings in the 0.01 nT to 0.1 nT range can be obtained. Variations of about 0.1 nT can be detected.

For hand/backpack carried units, PPM sample rates are typically limited to less than one sample per second. Measurements are typically taken with the sensor held at fixed locations at perhaps 10 meter increments.

The two main sources of measurement errors are magnetic impurities in the sensor, errors in the measurement of the frequency and ferrous material on the operator and in the instruments. If the sensor is rotated as the measurement is made, an additional error is generated.

Portable instruments are also limited by sensor volume (weight) and power consumption. PPMs work in field gradients up to 3,000nT/m which is adequate from most mineral exploration work. For higher gradient tolerance such as mapping banded iron formations and detecting large ferrous objects Overhauser magnetometers can handle 10,000nT/m and Cesium magnetometers can handle 30,000nT/m.

They are relatively inexpensive (< $US 8,000) and once widely used in mineral exploration. Three manufacturers dominate the market : GEM Systems, Geometrics and Scintrex. Popular models include G-856, Smartmag and GSM-18 and GSM-19T.

For mineral exploration they have been superseded by Overhauser and Cesium instruments which are both fast-cycling; the operator does not need to pause between readings, thereby increasing production.

Overhauser effect Magnetometer

The *Overhauser effect magnetometer* or *Overhauser magnetometer* measures the same fundamental effect as the *proton precession magnetometer*. By adding free radicals to the measurement fluid the nuclear Overhauser effect can be exploited to significantly improve upon the *proton precession magnetometer*. Rather than aligning the protons using a solenoid, a low power radio-frequency field is used to align (polarise) the electron spin of the free radicals which then couples to the protons via the Overhauser effect. This has two main advantages: driving the RF field takes a fraction of the energy (allowing lighter-weight batteries for portable units), and faster sampling as the electron-proton coupling can happen even as measurements are being taken. An Overhauser magnetometer produce readings with a 0.01nT to 0.02nT standard deviation while sampling once per second.

Fluxgate Magnetometer

Fluxgate magnetometers were invented in the 1930s by Victor Vacquier at Gulf Research Laboratories; Vacquier applied them during World War II as an instrument for detecting submarines, and after the war confirmed the theory of plate tectonics by using them to measure shifts in the magnetic patterns on the sea floor.

A fluxgate magnetometer consists of a small, magnetically susceptible, core wrapped by two coils of wire. An alternating electrical current is passed through one coil, driving the core through an alternating cycle of magnetic saturation; i.e., magnetised, unmagnetised, inversely magnetised, unmagnetised, magnetised, etc. This constantly changing field induces an electrical current in the second coil, and this output current is measured by a detector. In a magnetically neutral background, the input and output currents will match. However, when the core is exposed to a background field, it will be more easily saturated in alignment with that field and less easily saturated in opposition to it. Hence the alternating magnetic field, and the induced output current, will be out of step with the input current. The extent to which this is the case will depend on the strength of the background magnetic field. Often, the current in the output coil is integrated, yielding an output analog voltage, proportional to the magnetic field.

Fluxgate magnetometers, paired in a gradiometer configuration, are commonly used for archaeological prospecting and UXO detection such as the German military's popular Forster.

A wide variety of sensors are currently available and used to measure magnetic fields. Fluxgate magnetometers and gradiometers measure the direction and magnitude of magnetic fields. Fluxgates are affordable, rugged and compact. This, plus their typically low power consumption makes them ideal for a variety of sensing applications.

The typical fluxgate magnetometer consists of a "sense" (secondary) coil surrounding an inner "drive" (primary) coil that is wound around permeable core material. Each sensor has magnetic core elements that can be viewed as two carefully matched halves. An alternating current is applied to the drive winding, which drives the core into plus and minus saturation. The instantaneous drive current in each core half is driven in opposite polarity with respect to any external magnetic field.

In the absence of any external magnetic field, the flux in one core half cancels that in the other and the total flux seen by the sense coil is zero. If an external magnetic field is now applied, it will, at a given instance in time, aid the flux in one core half and oppose flux in the other. This causes a net flux imbalance between the halves, so that they no longer cancel one another. Current pulses are now induced in the sense winding on every drive current phase reversal (or at the 2nd, and all even harmonics). This results in a signal that is dependent on both the external field magnitude and polarity.

There are additional factors that affect the size of the resultant signal. These factors include the number of turns in the sense winding, magnetic permeability of the core, sensor geometry and the gated flux rate of change with respect to time. Phase synchronous detection is used to convert these harmonic signals to a DC voltage proportional to the external magnetic field.

Caesium Vapour Magnetometer

A basic example of the workings of a magnetometer may be given by discussing the common *optically pumped caesium vapour magnetometer* which is a highly sensitive (300 $fT/Hz^{0.5}$) and accurate device used in a wide range of applications. Although it relies on some interesting quantum mechanics to operate, its basic principles are easily explained.

The device broadly consists of a photon emitter containing a caesium light emitter or lamp, an absorption chamber containing caesium vapour and a "buffer gas" through which the emitted photons pass, and a photon detector, arranged in that order.

Polarization

The basic principle that allows the device to operate is the fact that a caesium atom can exist in any of nine energy levels, which is the placement of electron atomic orbitals around the atomic nucleus. When a caesium atom within the chamber encounters a photon from the lamp, it jumps to a higher energy state and then re-emits a photon and falls to an indeterminate lower energy state. The caesium atom is 'sensitive' to the photons from the lamp in three of its nine energy states, and therefore eventually, assuming a closed system, all the atoms will fall into a state in which all the photons from the lamp will pass through unhindered and be measured by the photon detector. At this point the sample (or population) is said to be polarized and ready for measurement to take place. This process is done continuously during operation.

Detection

Given that this theoretically perfect magnetometer is now functional, it can now begin to make measurements.

In the most common type of caesium magnetometer, a very small AC magnetic field is applied to the cell. Since the difference in the energy levels of the electrons is determined by the external magnetic field, there is a frequency at which this small AC field will cause the electrons to change states. In this new state, the electron will once again be able to absorb a photon of light. This causes a signal on a photo detector that measures the light passing through the cell. The associated electronics uses this fact to create a signal exactly at the frequency which corresponds to the external field.

Another type of caesium magnetometer modulates the light applied to the cell. This is referred to as a Bell-Bloom magnetometer after the two scientists who first investigated the effect. If the light is turned on and off at the frequency corresponding to the Earth's field, there is a change in the signal seen at the photo detector. Again, the associated electronics uses this to create a signal exactly at the frequency which corresponds to the external field.

Both methods lead to high performance magnetometers.

Applications

The caesium magnetometer is typically used where a higher performance magnetometer than the proton magnetometer is needed. In archaeology and geophysics, where the sensor is moved through an area and many accurate magnetic field measurements are needed, the caesium magnetometer has advantages over the proton magnetometer.

The caesium magnetometer's faster measurement rate allows the sensor to be moved through the area more quickly for a given number of data points. Caesium magnetometers are insensitive to rotation of the sensor while the measurement is being made.

The lower noise of the caesium magnetometer allows those measurements to more accurately show the variations in the field with position.

Spin-exchange Relaxation-free (SERF) Atomic Magnetometers

At sufficiently high atomic density, extremely high sensitivity can be achieved. Spin-exchange-relaxation-free (SERF) atomic magnetometers containing potassium, caesium or rubidium vapour operate similarly to the caesium magnetometers described above yet can reach sensitivities lower than 1 $fT/Hz^{0.5}$.

The SERF magnetometers only operate in small magnetic fields. The Earth's field is about 50 μT. SERF magnetometers operate in fields less than 0.5 μT.

As shown in large volume detectors have achieved 200 aT/Hz$^{0.5}$ sensitivity. This technology has greater sensitivity per unit volume than SQUID detectors.

The technology can also produce very small magnetometers that may in the future replace coils for detecting changing magnetic fields.

Rapid developments are ongoing in this area. This technology may produce a magnetic sensor that has all of its input and output signals in the form of light on fiber-optic cables. This would allow the magnetic measurement to be made in places where high electrical voltages exist.

SQUID Magnetometer

SQUIDs, or superconducting quantum interference devices, measure extremely small magnetic fields; they are very sensitive vector magnetometers, with noise levels as low as 3 fT/Hz$^{0.5}$ in commercial instruments and 0.4 fT/Hz$^{0.5}$ in experimental devices. Many liquid-helium-cooled commercial SQUIDs achieve a flat noise spectrum from near DC (less than 1 Hz) to tens of kilohertz, making such devices ideal for time-domain biomagnetic signal measurements. SERF atomic magnetometer demonstrated in a laboratory so far reaches competitive noise floor but in relatively small frequency ranges.

SQUID magnetometers require cooling with liquid helium (4.2 K) or liquid nitrogen (77 K) to operate, hence the packaging requirements to use them are rather stringent both from a thermal-mechanical as well as magnetic standpoint. SQUID magnetometers are most commonly used to measure the magnetic fields produced by brain or heart activity (magnetoencephalography and magnetocardiography, respectively). Geophysical surveys use SQUIDS from time to time, but the logistics is much more complicated than coil-based magnetometers.

Magnetic Surveys

Systematic surveys may be used to cover areas of interest such as exploring for mineral deposits or locating lost objects. Such surveys can be divided into

- Aeromagnetic survey
- Borehole
- Ground
- Marine.

Gradiometer

Magnetic gradiometers are pairs of magnetometers with their sensors separated by a fixed distance, usually horizontally. The readings are subtracted in order to measure the difference between the sensed magnetic fields, which measures the field gradients caused by magnetic anomalies.

This is one way of compensating both for the variability in time of the Earth's magnetic field and for other sources of electromagnetic interference, allowing more sensitive detection of anomalies. Because nearly equal values are being subtracted, the noise performance requirements for the magnetometers is more extreme. For this reason, high performance magnetometers are the rule in this type of system.

Gradiometers enhance shallow magnetic anomalies and are thus good for archaeological and some site investigation work. They are also good for real-time work such as Unexploded ordnance location. In the commercial world, it is twice as efficient to run a base station and use two (or more) mobile sensors to read parallel lines simultaneously (assuming data is stored and post-processed). In this manner both along-line and cross-line gradients can be calculated.

Position Control of Magnetic Surveys

In traditional mineral exploration and archaeological work, grid pegs placed by theodolite and tape measure were used to define the survey area. Some UXO surveys used ropes to define the lanes. Airborne surveys used radio triangulation beacons such a Siledus. Non-magnetic electronic hipchain triggers were developed to trigger magnetometers. There used rotary shaft encoders to measure distance along disposable cotton reels. Modern explorers use a range of low-magnetic signature GPS units including Real-Time Kinematic GPS.

Heading Errors in Magnetic Surveys

Magnetic surveys can suffer noise from a range of sources. Different magnetometer technologies suffer different kinds of noise problems. Heading errors are one group of noise. They comprise three sources :

- Sensor
- Console
- Operator.

Some total field sensors give different readings depending on their orientation. Magnetic materials in the sensor its self are the primary cause of this error.

In some magnetometers such as the vapour magnetometers (caesium potassium etc.) there are sources of heading error in the physics that contribute small amounts to the total heading error. Console noise comes from magnetic components on or within the console. These include ferrite in cores in inductors and transformers, steel frames around LCD's, legs on IC chips and steel cases in disposable batteries. Some popular MIL spec connectors also have steel springs.

Operators must take care to be magnetically clean and should check the 'magnetic hygiene' of all apparel and items carries during a survey. Acubra hats are very popular in Australia, however their steel rims must be removed before use on magnetic surveys. Steel rings on notepads, steel capped boots, steel springs in overall eyelets can all cause unnecessary noise in surveys. Pens, mobile phones and stainless steel implants can also be problematic.

The magnetic response (noise) from ferrous object on the operator and console can change with heading direction because of induction and remanence. Aeromagnetic survey aircraft and quad bike systems can use special compensators to correct for heading error noise.

Heading errors look like herringbone patterns in survey images. Alternate lines can also be corrugated.

Image Procesing of Magnetic Data

Recording data and image processing is a superior to real time work because subtle anomalies often missed by the operator (especially in magnetically noisy areas) can be correlated between lines, shapes and clusters better defined. A range of sophisticated enhancement techniques can also be used. There is also a hard copy and need for systematic coverage.

Early Magnetometers

In 1833, Carl Friedrich Gauss, head of the Geomagnetic Observatory in Göttingen, published a paper on measurement of the Earth's magnetic field. It described a new instrument that Gauss called a "magnometer" (a term which is still occasionally used instead of *magnetometer*). It consisted of a permanent bar magnet suspended horizontally from a gold fibre. A magnetometer may also be called a *gaussmeter*.

Petroleum Geologist

A petroleum geologist is an occupation that involves all aspects of oil discovery and production in the field of petroleum geology. Petroleum geologists are usually linked to the actual discovery of oil and the identification of possible oil deposits or leads. It can be a very labour intensive task involving several different fields of science and elaborate machinery. Petroleum geologists look at the structural and sedimentary aspects of the stratum/strata to identify possible oil traps.

Profillle

Petroleum geologists make the decision on where the location of drilling will occur. This is done by locating prospects within a sedimentary basin. Petroleum geologists determine a prospect's viability looking at seven main aspects, according to the orthodox view of petroleum exploration:

- Source- The presence of an organic-rich source rock capable of generating hydrocarbons during deep burial.
- Reservoir- The porous rock unit that collects the hydrocarbons expelled from the source rock and holds them inside a trap.
- Seal- The rock unit that inhibits the oil or gas from escaping vertically out of a hydrocarbon bearing reservoir rock.
- Trap- Structural or stratigraphic feature that captures migrating hydrocarbons into an economically producible accumulation.
- Timing- Geologic events must occur in a certain order e.g. that the trap formed before migration rather than after.
- Maturation- The process of thermal maturing of a source rock, leading to the cracking of its organic matter into oil and gas.
- Migration- The movement of the (less dense) oil or gas from the source rock into a reservoir rock and then into a trap.

These seven key aspects allow the Petroleum geologist to obtain a 1-dimensional idea of the subsurface. Some other data may be obtained via Geophysical methods. Geophysical logs show the seismology data of elastic waves, mainly seismic reflection. This allows a 3-dimensional look of the trap, and source rock. More data may be obtained from the mudlogger, who analyses the drill cuttings and the rock formation thicknesses.

Paleomagnetism

Paleomagnetism is the study of the record of the Earth's magnetic field in rocks. Certain minerals in rocks lock-in a record of the direction and intensity of the magnetic field when they form. This record provides information on the past behaviour of Earth's magnetic field and the past location of tectonic plates.

The record of geomagnetic reversals preserved in volcanic and sedimentary rock sequences (magnetostratigraphy) provides a time-scale that is used as a geochronologic tool. Geophysicists who specialize in paleomagnetism are called paleomagnetists.

Paleomagnetists led the revival of the theory of continental drift and its transformation into plate tectonics. Apparent polar wander paths provided the first clear geophysical evidence for continental drift, while marine magnetic anomalies did the same for seafloor spreading. Paleomagnetism continues to extend the history of plate tectonics back in time and are applied to the movement of continental fragments, or terranes.

Paleomagnetism relied heavily on new developments in rock magnetism, which in turn has provided the foundation for new applications of magnetism. These include biomagnetism, magnetic fabrics (used as strain indicators in rocks and soils), and environmental magnetism.

History

As early as the 18th century it was noticed that compass needles deviated near strongly magnetized outcrops. In 1797, Von Humboldt attributed this magnetization to lightning strikes (and lightning strikes do often magnetize surface rocks). In the 19th century studies of the direction of magnetization in rocks showed that some recent lavas were magnetized parallel to the Earth's magnetic field. Early in the 19th century, work by David, Brunhes and Mercanton showed that many rocks were magnetized antiparallel to the field. Motonori Matuyama showed that the Earth's magnetic field reversed in the mid-Quaternary, a reversal now known as the Brunhes-Matuyama reversal.

The British physicist P.M.S. Blackett provided a major impetus to paleomagnetism by inventing a sensitive astatic magnetometer in 1956. His intent was to test his theory that the geomagnetic field was related to the Earth's rotation, a theory that he ultimately rejected; but the astatic magnetometer became the basic tool of paleomagnetism and led to a revival of the theory of continental drift. Alfred Wegener first proposed in 1915 that continents had once been joined together and had since moved apart. Although he produced an abundance of circumstantial evidence, his theory met with little acceptance for two reasons:

(1) no mechanism for continental drift was known, and

(2) there was no way to reconstruct the movements of the continents over time.

Keith Runcorn and Edward A. Irving constructed apparent polar wander paths for Europe and North America. These curves diverged, but could be reconciled if it was assumed that the continents had been in contact up to 200 million years ago. This provided the first clear geophysical evidence for continental drift. Then in 1963, Morley, Vine and Matthews showed that marine magnetic anomalies provided evidence for seafloor spreading.

Fields of Paleomagnetism

Paleomagnetism is studied on a number of scales:

- *Secular variation studies* look at small-scale changes in the direction and intensity of the Earth's magnetic field. The magnetic north pole is constantly shifting relative to the axis of rotation of the Earth. Magnetism is a vector and so magnetic field variation is made up of palaeodirectional measurements of magnetic declination and magnetic inclination and palaeointensity measurements.
- *Magnetostratigraphy* uses the polarity reversal history of the Earth's magnetic field recorded in rocks to determine the age of those rocks. *Reversals* have occurred at irregular intervals throughout Earth history. The age and pattern of these reversals is known from the study of sea floor spreading zones and the dating of volcanic rocks.

Principles of Remanent Magnetization

The study of paleomagnetism is possible because iron-bearing minerals such as magnetite may record past directions of the Earth's magnetic field. Paleomagnetic signatures in rocks can be recorded by three different mechanisms.

Thermoremanent Magnetization

Iron-titanium oxide minerals in basalt and other igneous rocks may preserve the direction of the Earth's magnetic field when the rocks cool through the Curie temperatures of those minerals. The Curie temperature of magnetite, a spinel-group iron oxide, is about 580°C, whereas most basalt and gabbro are completely crystallized at temperatures above 900°C. Hence, the mineral grains are not rotated physically to align with the Earth's field, but rather they may record the orientation of that field. The record so preserved is called a thermoremanent magnetization (TRM). Because complex oxidation reactions may occur as igneous rocks cool after crystallization, the orientations of the Earth's magnetic field are not always accurately recorded, nor is the record necessarily maintained. Nonetheless, the record has been preserved well enough in basalts of the ocean crust to have been critical in the development of theories of sea floor spreading related to plate tectonics. TRM can also be recorded in pottery kilns, hearths, and burned adobe buildings. The discipline based on the study of thermoremanent magnetisation in archaeological materials is called archaeomagnetic dating.

Detrital Remanent Magnetization

In a completely different process, magnetic grains in sediments may align with the magnetic field during or soon after deposition; this is known as *detrital remanent magnetization* (DRM). If the magnetization is acquired as the grains are deposited, the result is a depositional detrital remanent magnetization (dDRM); if it is acquired soon after deposition, it is a post-depositional detrital remanent magnetization (pDRM).

Chemical Remanent Magnetization

In a third process, magnetic grains grow during chemical reactions, and record the direction of the magnetic field at the time of their formation. The field is said to be recorded by *chemical remanent magnetization* (CRM). A common form of chemical remanent magnetization is held by the mineral hematite, another iron oxide. Hematite forms through chemical oxidation reactions of other minerals in the rock including magnetite. Redbeds, clastic sedimentary rocks (such as sandstones) are red because of hematite that formed during sedimentary diagenesis. The CRM signatures in redbeds can be quite useful and they are common targets in magnetostratigraphy studies.

Isothermal Remanent Magnetization

Remanence that is acquired at a fixed temperature is called *isothermal remanent magnetization (IRM)*. Remanence of this sort is not useful for paleomagnetism, but it can be acquired as a result of lightning strikes. Lightning-induced remanent magnetization can be distinguished by its high intensity and rapid variation in direction over scales of centimeters.

IRM is often induced in drill cores by the magnetic field of the steel core barrel. This contaminant is generally parallel to the barrel, and most of it can be removed by heating up to about 400 ! or demagnetizing in a small alternating field.

In the laboratory, IRM is induced by applying fields of various strengths and is used for many purposes in rock magnetism.

Paleomagnetic Procedure

Collecting Samples on Land

Paleomagnetists, like many geologists, gravitate towards outcrops because layers of rock are exposed. Road cuts are a convenient man-made source of outcrops.

> *"And everywhere, in profusion along this half mile of [roadcut], there are small, neatly cored holes... appears to be a Hilton for wrens and purple martins."*

There are two main goals of sampling:

1. Retrieve samples with accurate orientations, and
2. Reduce statistical uncertainty.

One way to achieve the first goal is to use a rock coring drill that has a pipe tipped with diamond bits. The drill cuts a cylindrical space around some rock. This can be messy - the drill must be cooled with water, and the result is mud spewing out of the hole. Into this space is inserted another pipe with compass and inclinometer attached. These provide the orientations. Before this device is removed, a mark is scratched on the sample. After the sample is broken off, the mark can be augmented for clarity.

Examples

Paleomagnetic evidence, both reversals and polar wandering data, was instrumental in verifying the theories of continental drift and plate tectonics in the 1960s and 70s. Some applications of paleomagnetic evidence to reconstructing histories of terranes have continued to arouse controversies. Paleomagnetic evidence also is used in constraining possible ages for rocks and processes and in reconstructions of the deformational histories of parts of the crust.

Reversal magnetostratigraphy is often used to estimate the age of fossil and hominin bearing sites.

Paleomagnetic studies are combined with geochronological methods to determine absolute ages for rocks in which the magnetic record is preserved. For igneous rocks such as basalt, commonly used methods include potassium-argon and argon-argon geochronology.

Concession (Contract)

A concession is a business operated under a contract or license associated with a degree of exclusivity in business within a certain geographical area. For example, sports arenas or public parks may have concession stands. Many department stores contain numerous concessions operated by other retailers. Similarly, public services such as water supply may be operated as concessions.

There are now over 3000 garden centre concessions. These concessions are operated from fixed internal areas and are fitted out by the operating concession company. Concessions are either unmanned or manned formats. Short term concessions are operated on garden centres and shopping malls from promotional space for short periods of between a single day to six months

The owner of the concession — the *concessionaire* — pays either a fixed sum or a percentage of revenue to the entity with the ability to assign exclusive rights for an area or facility. A concession may involve the transfer to the concessionaire of the right to use some existing infrastructure required to carry out a business (such as a water supply system in a city); in some cases, such as mining, it may involve merely the transfer of exclusive or non-exclusive easements.

In the case of a public service concession, a private company enters into an agreement with the government to have the exclusive right to operate, maintain and carry out investment in a public utility (such as a water supply system) for a given number of years. Other forms of contracts between public and private entities, namely lease contract and management contract (in the water sector often called by the French term *affermage*), are closely related but differ from a concession in the rights of the operator and its remuneration.

A lease gives a company the right to operate and maintain a public utility, but investment remains the responsibility of the public. Under a management contract the operator will collect the revenue only on behalf of the government and will in turn be paid an agreed fee.

Some countries, such as Chile, Canada and the United Kingdom, have built toll highways on a concession basis (e.g. M6 Toll in the UK). Concessions have also been used in London in construction of extensions to the Docklands Light Railway system.

Public-private partnerships (PPPs) and private finance initiatives (PFIs) may be considered similar to concessions.

Production Sharing Agreement

Production sharing agreements (PSAs) are a common type of contract signed between a government and a resource extraction company (or group of companies) concerning how much of the resource (usually oil) extracted from the country each will receive.

Production sharing agreements were first used in Bolivia in the early 1950s, although their first implementation similar to today's was in Indonesia in the 1960s. Today they are often used in the Middle East and Central Asia. In production sharing agreements the country's government awards the execution of exploration and production activities to an oil company.

The oil company bears the mineral and financial risk of the initiative and explores, develops and ultimately produces the field as required. When successful, the company is permitted to use the money from produced oil to recover capital and operational expenditures, known as "cost oil".

The remaining money is known as "profit oil", and is split between the government and the company, typically at a rate of about 80% for the government, 20% for the company. In some production sharing agreements, changes in international oil prices or production rate can affect the company's share of production.

Production sharing agreements can be beneficial to governments of countries that lack the expertise and/or capital to develop their resources and wish to attract foreign companies to do so. They can be very profitable agreements for the oil companies involved, but often involve considerable risk.

Pumpjack

A pumpjack (nodding donkey, pumping unit, horsehead pump, beam pump, sucker rod pump (SRP), grasshopper pump, thirsty bird, jack pump) is the overground drive for a reciprocating piston pump in an oil well. It is used to mechanically lift liquid out of the well if there is not enough bottom hole pressure for the liquid to flow all the way to the surface. The arrangement is commonly used for onshore wells producing little oil. Pumpjacks are common in oil-rich areas. Depending on the size of the pump, it generally produces 5 to 40 litres of liquid at each stroke. Often this is an emulsion of crude oil and water. Pump size is also determined by the depth and weight of the oil to remove, with deeper extraction requiring more power to move the heavier lengths of sucker rods.

A pumpjack converts the rotary mechanism of the motor to a vertical reciprocating motion to drive the pump shaft, and is exhibited in the characteristic nodding motion. The engineering term for this type of mechanism is a walking beam. It was often employed in stationary and marine steam engine designs in the 18th and 19th centuries.

Above Ground

In the early days, pumpjacks were actuated by rod lines running horizontally above the ground to an eccentric wheel in a mechanism known as a Central Power. The Central Power itself, which might operate a dozen or more pumpjacks, was powered by a steam or internal combustion engine or by an electric motor. Among the difficulties with this scheme was keeping the system balanced as individual well loads changed.

Modern pumpjacks are powered by a prime mover. This is commonly an electric motor, but combustion engines are used in isolated locations without economic access to electricity. Common off-grid pumpjack engines run on casing gas produced from the well, but pumpjacks have been run on many types of fuel, such as propane and diesel. In harsh climates such motors and engines may be housed inside a shack to protect them from the elements.

The prime mover of the pumpjack runs a set of pulleys to the transmission which in turn drives a pair of cranks, generally with counterweights on them to assist the motor in lifting the heavy string of rods. The cranks in turn raise and lower one end of an I-beam which is free to move on an A-frame.

On the other end of the beam, there is a curved metal box called a Horse Head or Donkeys Head, named so due to its appearance. A cable made of steel (or, occasionally, fiberglass) called a bridle, connects the horse head to the polished rod, a piston that passes through the stuffing box. The polished rod has a close fit to the stuffing box, letting it move in and out of the tubing without fluid escaping. (The tubing is a pipe that runs to the bottom of the well through which the liquid is produced.) The bridle follows the curve of the horse head as it lowers and raises to create an almost completely vertical stroke. The polished rod is connected to a long string of rods called sucker rods, which run through the tubing all the way to the down-hole pump, usually positioned near the bottom of the well.

Down-hole

At the bottom of the tubing is the down-hole pump. This pump has two ball check valves: a stationary valve at bottom called the standing valve, and a valve on the piston connected to the bottom of the sucker rods that travels up and down as the rods reciprocate, known as the travelling valve. Reservoir fluid enters from the formation into the bottom of the borehole through perforations that have been made through the casing and cement (the casing is a larger metal pipe that runs the length of the well, which has cement placed between it and the earth; the tubing, pump and sucker rods are all inside the casing).

When the rods at the pump end are travelling up, the travelling valve is closed and the standing valve is open (due to the drop in pressure in the pump barrel). Consequently, the pump barrel fills with the fluid from the formation as the travelling piston lifts the previous contents of the barrel upwards. When the rods begin pushing down, the travelling valve opens and the standing valve closes (due to an increase in pressure in the pump barrel). The travelling valve drops through the fluid in the barrel (which had been sucked in during the upstroke). The piston then reaches the end of its stroke and begins its path upwards again, repeating the process.

Often, gas is produced through the same perforations as the oil. This can be problematic if gas enters the pump, because it can result in what is known as gas locking, where insufficient pressure builds up in the pump barrel to open the valves (due to compression of the gas) and little or nothing is pumped.

To preclude this, the inlet for the pump can be placed below the perforations. As the gas-laden fluid enters the well bore through the perforations, the gas bubbles up the annulus (the space between the casing and the tubing) while the liquid moves down to the standing valve inlet. Once at the surface, the gas is collected through piping connected to the annulus.

Water Well Pumpjacks

Pumpjacks can also be used to drive what would now be considered old-fashioned hand-pumped water wells. The scale of the technology is smaller than for an oil well, and can typically fit on top of an existing hand-pumped well head. The technology is simple, typically using a parallel-bar double-cam lift driven from a very low horsepower electric motor.

Although the flow rate for a water well pumpjack is low compared to a jet pump and the lifted water is not pressurized, the water well pumpjack has the option of hand pumping in an emergency, by hand-rotating the pumpjack cam to its lowest position, and attaching a manual handle to the top of the wellhead rod.

Gas Lift

Gas lift is one of a number of processes used to artificially lift oil or water from wells where there is insufficient reservoir pressure to produce the well. The process involves injecting gas through the tubing-casing annulus. Injected gas aerates the fluid to reduce its density; the formation pressure is then able to lift the oil column and forces the fluid out of the wellbore. Gas may be injected continuously or intermittently, depending on the producing characteristics of the well and the arrangement of the gas-lift equipment.

The amount of gas to be injected to maximize oil production varies based on well conditions and geometries. Too much or too little injected gas will result in less than maximum production. Generally, the optimal amount of injected gas is determined by well tests, where the rate of injection is varied and liquid production (oil and perhaps water) is measured.

Although the gas is recovered from the oil at a later separation stage, the process requires energy to drive a compressor in order to raise the pressure of the gas to a level where it can be re-injected.

The gas-lift mandrel is a device installed in the tubing string of a gas-lift well onto which or into which a gas-lift valve is fitted. There are two common types of mandrels. In a conventional gas-lift mandrel, a gas-lift valve is installed as the tubing is placed in the well. Thus, to replace or repair the valve, the tubing string must be pulled. In the side-pocket mandrel, however, the valve is installed and removed by wireline while the mandrel is still in the well, eliminating the need to pull the tubing to repair or replace the valve.

A gas-lift valve is a device installed on (or in) a gas-lift mandrel, which in turn is put on the production tubing of a gas-lift well. Tubing and casing pressures cause the valve to open and close, thus allowing gas to be injected into the fluid in the tubing to cause the fluid to rise to the surface. In the lexicon of the industry, gas-lift mandrels are said to be "tubing retrievable" wherein they are deployed and retrieved attached to the production tubing.

Chapter 5

Petroleum Production Engineering

Production engineering is a subset of petroleum engineering. Production engineers design and select equipment to extract and treat oil and gas well fluids. They often are degreed as petroleum engineers, although they may come from other technical disciplines (e.g., mechanical engineering) and subsequently be trained by an oil and gas company.

Production engineers' responsibilities include:

1. Evaluating inflow and outflow performance between the reservoir and the wellbore.
2. Designing completion systems, including tubing selection, perforating, sand control, matrix stimulation, and hydraulic fracturing.
3. Selecting artificial lift equipment, including sucker-rod lift (typically beam pumping), gas lift, electrical submersible pumps, subsurface hydraulic pumps, progressing-cavity pumps, and plunger lift.
4. Selecting equipment for surface facilities that separate and measure the produced fluids (oil, natural gas, water, and impurities), prepare the oil and gas for transportation to market, and handle disposal of any water and impurities.

Subsurface Engineer

Subsurface engineers (also known as Completion engineers) are a subset within Petroleum Engineering and typically work closely with Drilling engineers. The job of a Subsurface Engineer is to effectively select equipment that will best suit the subsurface environment in order to best produce the hydrocarbon reserves. Once the hardware has been selected, a Subsurface Engineer will monitor and adjust the equipment to ensure the well and reservoir produces under ideal circumstances.

Subsurface engineers must design a successful well completion system by selecting equipment that is adequate for both downhole environments and applications. Considerations must be given to the various functions under which the completion equipment must operate and the effects any changes in temperatures or differential pressure will have on the equipment.

The completion system must also be efficient and cost effective to achieve maximum production and financial goals. Another factor in the selection of specific completion equipment is the production rates of the well. The typical job duties of a Subsurface engineer include managing the interface between the reservoir and the well, including perforations, sand control, artificial lift, downhole flow control, and downhole monitoring equipment.

Additional responsibilities of a Subsurface engineer include: performing a cost and risk analysis on the design, contacting vendors for the rental, purchase, and shipment of equipment, and working closely with fellow employees (geologists, reservoir engineers, drilling engineers, and production engineers).

Design Components

The design components considered to perform a well completion may include:

- Cost and risk analysis
- Determining the specifications for the wellbore clean-out
- Use of specific Packer assemblies
- Determining specific tool selection to operate equipment within the well
- Asses possible equipment load specifications and incorporation of safety factors
- Best use of flow control accessories (sliding sleeves and safety valves)
- Determining the appropriate perforating shots per foot and charges based on the target formations
- Acidifying the formation to inhibit flow of hydrocarbons
- Sand Control operations to increase production

- Prevention of formation sand production with the use of wire screens
- Review Well logs to determine equipment placement within the well
- Determination of specific production pipe regarding well flow rates
- Selection of equipment to maintain well stability
- Oversee completion operations.

Artificial Lift

Artificial lift refers to the use of artificial means to increase the flow of liquids, such as crude oil or water, from a production well. Generally this is achieved by the use of a mechanical device inside the well (pump or velocity string) or by decreasing the weight of the hydrostatic column by injecting gas into the liquid some distance down the well. Artificial lift is needed in wells when there is insufficient pressure in the reservoir to lift the produced fluids to the surface, but often used in naturally flowing wells (which do not technically need it) to increase the flow rate above what would flow naturally. The produced fluid can be oil and/or water, typically with some amount of gas included.

Why use Artificial Lift

Any liquid-producing reservoir will have a 'reservoir pressure': some level of energy or potential that will force fluid (liquid and/or gas) to areas of lower energy or potential. You can think of this much like the water pressure in your municipal water system. As soon as the pressure inside a production well is decreased below the reservoir pressure, the reservoir will act to fill the well back up, just like opening a valve on your water system. Depending on the depth of the reservoir (deeper results in higher pressure requirement) and density of the fluid (heavier mixture results in higher requirement), the reservoir may or may not have enough potential to push the fluid to the surface.

Most oil production reservoirs have sufficient potential to produce oil and gas - which are light - naturally in the early phases of production. Eventually, as water - which is heavier than oil and much heavier than gas - encroaches into production and reservoir pressure decreases as the reservoir depletes, all wells will stop flowing naturally. At some point, most well operators will implement an artificial lift plan to continue and/or to increase production. Most water-producing wells, by contrast, will need artificial lift from the very beginning of production because they do not benefit from the lighter density of oil and gas.

Artificial Lift Technologies

Hydraulic Pumping Systems

Hydraulic pumping systems transmit energy to the bottom of the well by means of pressurized power fluid that flows down in the wellbore tubular to a subsurface pump. There are two types of hydraulic subsurface pump:

a) a reciprocating piston pump, where one side is powered by the injected fluid while the other side pumps the produced fluids to surface, and

b) a jet pump, where the injected fluid passes through a nozzle creating a venturi effect pushing the produced fluids to surface.

These systems are very versatile and have been used in shallow depths (1000 ft) to deeper wells (18,000 ft), low rate wells with production in the tens of barrels per day to wells producing in excess of 10,000 barrels (1,600 m^3) per day. Certain substances can be mixed in with the injected fluid to help deal or control with corrosion, paraffin and emulsion problems. Hydraulic pumping systems are also suitable for deviated wells where conventional pumps such as the rod pump are not feasible.

These systems have also some disadvantages. They are sensitive to solids and are the least efficient lift method. While typically the cost of deploying these systems has been very high, new coiled tubing umbilical technologies are in some cases greatly reducing the cost.

ESP

Electric Submersible Pumps consist of a) a downhole pump, which is a series of centrifugal pumps, b) a separator or protector, which function is to prevent that produced fluids enter the electrical motor, c) the electrical motor, which transforms the electrical power into kinetic energy to turn the pump, and d) an electric power cable that connects the motor to the surface control panel. ESP is a very versatile artificial lift method and can be found in operating environments all over the world. They can handle a very wide range of flow rates (from 200 to 90,000 barrels (14,000 m^3) per day) and lift requirements (from virtually zero to 10,000 ft (3,000 m) of lift). They can be modified to handle contaminants commonly found in oil, aggressive corrosive fluids such as H_2S and CO_2, and exceptionally high downhole temperatures.

Increasing water cut has been shown to have no significant detrimental effect on the ESP performance. It is possible to locate them in vertical, deviated, or horizontal wells, but it is recommended to deploy them in a straight section of casing for optimum run life performance. Although latest developments are aimed to enhance the ESP capabilities to handle gas and sand, they still need more technological development to avoid gas locked and internal erosion.

Until recently, ESP's have come with an often prohibitive price tag due to the cost of deployment which can be in excess of $20,000.

Gas Lift

Gas Lift is another widely used artificial lift method. As the name denotes, gas is injected in the tubing to reduce the weight of the hydrostatic column, thus reducing the back pressure and allowing the reservoir pressure to push the mixture of produce fluids and gas up to the surface. The gas lift can be deployed in a wide range of well conditions (up to 30,000 bbl/d (4,800 m^3/d) and down to 15,000 ft (4,600 m)).

They handle very well abrasive elements and sand, and the cost of workover is minimum. The gas lifted wells are equipped with side pocket mandrel and gas lift injection valves. This arrangement allows a deeper gas injection in the tubing. The gas lift system has some disadvantages. There has to be a source of gas, some flow assurance problems such as hydrates can be triggered by the gas lift.

PCP

Progressing Cavity Pumps, PCP, are also widely applied in the oil industry. The PCP consists of a stator and a rotor. The rotor is rotated using either a top side motor or a bottom hole motor. The rotation created sequential cavities and the produced fluids are pushed to surface. The PCP is a flexible system with a wide range of applications in terms of rate(up to 5,000 bbl/d (790 m^3/d) and 6,000 ft (1,800 m)). They offer outstanding resistance to abrasives and solids but they are restricted to setting depths and temperatures. Some components of the produced fluids like aromatics can also deteriorate the stator's elastomer.

Rod Pumps

Rod Pumps are long slender cylinders with both fixed and moveable elements inside. The pump is designed to be inserted inside the tubing of a well and its main purpose is to gather fluids from beneath it and lift them to the surface.

The most important components are: the barrel, valves (travelling and fixed) and the piston. It also has another 18 to 30 components which are called "fittings".

Components

Every part of the pump is important for its correct operation. The most commonly used parts are described below:

- Barrel: The barrel is a long cylinder, which can be from 10 to 36 feet (11 m) long, with a diameter of 1.25 inches (32 mm) to 3.75 inches (95 mm). After experience with several materials for its construction, the API (American Petroleum Institute) standardized the use of two materials or compositions for this part: carbon steel and brass, both with an inside coating of chrome. The advantage of brass against the harder carbon steel is its 100% resistance to corrosion.
- Piston/Plunger: This is a nickel-metal sprayed steel cylinder that goes inside the barrel. Its main purpose is to create a sucking effect that lifts the fluids beneath it and then, with the help of the valves, take the fluids above it, progressively, out of the well. It achieves this with a reciprocating up and down movement.
- Valves: The valves have two components - the seat and the ball - which create a complete seal when closed. The most commonly used seats are made of carbon nitride and the ball is often made of silicon nitride. In the past, balls of iron, ceramic and titanium were used. Titanium balls are still being used but only where crude oil is extremely dense and/or the quantity of fluid to be lifted is large. The most common configuration of a rod pump requires two valves, called the travelling valve and the fixed (or static or standing) valve.
- Piston rod: This is a rod that connects the piston with the outside of the pump. Its main purpose is to transfer the up/down reciprocating energy produced by the "Nodding Donkey" (pumping unit) installed above ground.
- Fittings: The rest of the parts of the pump are called fittings and are, basically, small pieces designed to keep everything hold together in the right place. Most of these parts are designed to let the fluids pass uninterrupted.
- Filter/Strainer: The job of the filter, as guessed, is to stop big parts of rock, rubber or any other garbage that might be loose in the well from being sucked into the pump. There are several types of filters, with the most common being an iron cylinder with enough holes in it to permit the entrance of the amount of fluid the pump needs.

Seismic Source

A seismic source is a device that generates controlled seismic energy used to perform both reflection and refraction seismic surveys. A seismic source can be simple, such as dynamite, or it can use more sophisticated technology, such as a specialized air gun. Seismic sources can provide single pulses or continuous sweeps of energy. Both types of seismic sources generate seismic waves, which travel through a medium such as water or layers of rocks. Some of the waves then reflect and refract and is recorded by receivers, such as geophones or hydrophones.

Seismic sources may be used to investigate shallow subsoil structure, for engineering site characterisation, or to study deeper structures, usually in the search for petroleum or mineral deposits, or for scientific investigation. The returning signals from the sources are detected by seismic sensors (geophones or hydrophones), laid in known locations relative to the position of the source. The recorded signals are then subjected to specialist processing and interpretation to yield comprehensible data about the subsurface.

Source Model

A seismic source signal has the following characteristics:

1. generated as an impulsive source
2. band-limited
3. the generated waves are time-varying.

The generalized equation that shows all above properties is:

$$s(t) = \beta e^{-\alpha t^2} \sin(2\pi f_{max} t)$$

where f_{max} is the maximum frequency component of the generated waveform.

Types of Sources

Explosives

Explosives, such as dynamite, can be used as crude but effective sources of seismic energy. Generally the explosive charges are placed between 20 feet to 250 feet below ground. The charges are placed in a hole that is drilled with dedicated drilling equipment for this purpose. This type of seismic drilling is often referred to as "Shot Hole Drilling".

A common drill rig used for "Shot Hole Drilling" is the ARDCO C-1000 drill mounted on an ARDCO K 4X4 buggy. These drill rigs often use water or air in assisting the drilling.

Air Gun

An air gun is used for marine reflection and refraction surveys. It consists of one or more pneumatic chambers that are pressurized with compressed air at pressures from 2,000 pounds per square inch to 3,000 pounds per square inch (14 to 21 MPa). The air gun array is submerged below the water surface, and is towed behind a ship. When the air gun is fired, a solenoid is triggered, which releases air into a fire chamber which in turn causes a piston to move and thereby allowing the air to escape the main chamber and to produce a pulse of acoustic energy. Air gun arrays are built up of up to 48 individual air guns with different size chambers, the aim being to create the optimum initial shock wave with minimum reverberation of the bubble after the first shot.

Gun arrays can be fired in flip-flop mode; typically this would be 48 guns per source, which would be selected and fired alternately. Large chambers (i.e., greater than 70 cubic inches or 1.15 L) tend to give low frequency signals, and the small chambers (less than 70 cubic inches) give higher frequency signals. The air gun is made from the highest grades of corrosion resistant stainless steel.

Plasma Sound Source

Plasma sound source fired in small swimming pool. A plasma sound source (PSS), otherwise called a spark gap sound source, or simply a sparker, is a means of making very low frequency sonar pulse underwater.

Figure: *Plasma sound source fired in small swimming pool.*

For each firing, it stores electric charge in a large high-voltage bank of capacitors, and then releases all the stored energy in an arc across electrodes in the water. The underwater spark discharge produces a high-pressure plasma and vapour bubble, which expands and collapses, making a loud sound. Most of the sound produced is between 20 and 200 Hz.

The PSS has also been used for sonar. There are also plans to use PSS as a non-lethal weapon against submerged divers.

Thumper Truck

Dynamite was the only source used until 1953 when the weight dropping Thumper technique was introduced.

A thumper truck (or weight-drop) truck is a vehicle mounted ground impact which can used to provide the seismic source. A heavy weight is raised by a hoist at the back of the truck and dropped, possibly about three metres, to impact (or "thump") the ground.

To augment the signal, the weight may be dropped more than once at the same spot, the signal may also be increased by thumping at several nearby places in an array whose dimensions may be chosen to enhance the seismic signal by spatial filtering.

Thumping might be less damaging to the environment than firing explosives in shot-holes, though a heavily thumped seismic line with transverse ridges every few metres might create long-lasting disturbance of the soil. An advantage of the thumper (later shared with Vibroseis), especially in politically unstable areas, was that no explosives were required.

Thumper's advanced technology what we call AWD "Accelerated Weight Drop" seismic energy source, where a high pressure gas (min 1000 PSI) is used to accelerate a heavy weight Hammer (5000Kg) to hit a base plate coupled from a distance of 2m - 3m, which is coupled to the ground to generate a acoustic pulse. Several thumps were stacked to enhance signal to noise ratio.

The very basic assumption for choosing a seismic source are all satisfied by AWD as follows

1. Target Depth: Penetration of the required depth
2. Frequency Content of the seismic wavelet: Bandwidth for the required resolution
3. Strength: Signal to Noise ratio
4. Environment friendly
5. Economic: Availability and cost.

It has been estimated that approximately 1% of the chemical energy of a small charge of dynamite is converted into useful P wave energy of the seismic signal. It is unknown how inefficient mechanical impulsive sources are in converting primary energy into useful seismic energy.

To overcome these inefficiencies and obtain a higher amplitude of the seismic wavelet or increase the signal to noise ratio, it is usual to either

1) Increase the primary energy of the source (increase the dynamite from 1 to 10 kgs, drop a weight from 2 meters instead of 1 meter), and/or
2) synchronize 2 or more sources to fire simultaneously, and/or
3) stack records sequentially.

Each of these methods has trade-offs.

Electromagnetic Pulse Energy Source (Non-Explosive)

EMP sources based on the electrodynamic and electromagnetic principles. A Charge in the cylinder lifts the plate in EMP and released as soon as the it discharges and a pulse is generated of high energy. Look at the amplitude spectrum of the EMP source, the frequency content for dynamite is taken upto 125 Hz, but amplitude dies after 70 Hz. But the amplitude spectrum you can very well see it goes upto 90 hz. So its quite comparable.

Seismic Vibrator

A Seismic vibrator, commonly known by its trademark name Vibroseis, propagates energy signals into the Earth over an extended period of time as opposed to the near instantaneous energy provided by impulsive sources. The data recorded in this way must be *correlated* to convert the extended source signal into an impulse. The source signal using this method was originally generated by a servo-controlled hydraulic vibrator or *shaker unit* mounted on a mobile base unit, but electro-mechanical versions have also been developed.

Vibroseis was developed by the Continental Oil Company (Conoco) during the 1950s and was a trademark until the company's patent lapsed.

Boomer Sources

Boomer sound sources are used for shallow water seismic surveys, mostly for engineering survey applications. Boomers are towed in a floating sled behind a survey vessel.

Similarly to the plasma source, it stores energy in capacitors, but it discharges through a flat spiral coil instead of generating a spark. A copper plate adjacent to the coil flexes away from the coil as the capacitors are discharged. This flexing is transmitted into the water as the seismic pulse.

Originally the storage capacitors were placed in a steel container (the bang box) on the survey vessel. The high voltages used, typically 3,000 V, required heavy cables and strong safety containers. Recently, low voltage boomers have become available. These use capacitors on the towed sled, allowing efficient energy recovery, lower voltage power supplies and lighter cables. The low voltage systems are generally easier to deploy and have fewer safety concerns.

Noise Sources

Correlation-based processing techniques also enable seismologists to image the interior of the Earth at multiple scales using natural (e.g., the oceanic microseism) or artificial (e.g., urban) background noise as a seismic source. For example, under ideal conditions of uniform seismic illumination, the correlation of the noise signals between two seismographs provides an estimate of the bidirectional seismic impulse response.

Reservoir Simulation

Reservoir simulation is an area of reservoir engineering in which computer models are used to predict the flow of fluids (typically, oil, water, and gas) through porous media.

Uses

Reservoir simulation models are used by oil and gas companies in the development of new fields. Also, models are used in developed fields where production forecasts are needed to help make investment decisions. As building and maintaining a robust, reliable model of a field is often time-consuming and expensive, models are typically only constructed where large investment decisions are at stake. Improvements in simulation software have lowered the time to develop a model. Also, models can be run on personal computers rather than more expensive workstations.

For new fields, models may help development by identifying the number of wells required, the optimal completion of wells, the present and future needs for artificial lift, and the expected production of oil, water and gas.

For ongoing reservoir management, models may help in improved oil recovery by hydraulic fracturing. Highly deviated or horizontal wells can also be represented.

Specialized software may be used in the design of hydraulic fracturing, then the improvements in productivity can be included in the field model. Also, future improvement in oil recovery with pressure maintenance by re-injection of produced gas or by water injection into an aquifer can be evaluated. Water flooding resulting in the improved displacement of oil is commonly evaluated using reservoir simulation.

The application of enhanced oil recovery (EOR) processes requires that the field possesses the necessary characteristics to make application successful. Model studies can assist in this evaluation. EOR processes include miscible displacement by natural gas, CO_2, or nitrogen and chemical flooding (polymer, alkaline, surfactant, or a combination of these). Special features in simulation software is needed to represent these processes. In some miscible applications, the "smearing" of the flood front, also called numerical dispersion, may be a problem.

Reservoir simulation is used extensively to identify opportunities to increase oil production in heavy oil deposits. Oil recovery is improved by lowering the oil viscosity by injecting steam or hot water. Typical processes are steam soaks (steam is injected, then oil produced from the same well) and steam flooding (separate steam injectors and oil producers). These processes require simulators with special features to account for heat transfer to the fluids present and the formation, the subsequent property changes and heat losses outside of the formation.

A recent application of reservoir simulation is the modelling of coalbed methane (CBM) production. This application requires a specialized CBM simulator. In addition to the normal fractured (fissured) formation data, CBM simulation requires gas content data values at initial pressure, sorption isotherms, diffusion coefficient, and parameters to estimate the changes in absolute permeability as a function of pore-pressure depletion and gas desorption.

Fundamentals

Traditional finite difference simulators dominate both theoretical and practical work in reservoir simulation. Conventional FD simulation is underpinned by three physical concepts: conservation of mass, isothermal fluid phase behaviour, and the Darcy approximation of fluid flow through porous media. Thermal simulators (most commonly used for heavy oil applications) add conservation of energy to this list, allowing temperatures to change within the reservoir.

Numerical techniques and approaches that are common in modern simulators:

- Most modern FD simulation programs allow for construction of 3-D representations for use in either full-field or single-well models. 2-D approximations are also used in various conceptual models, such as cross-sections and 2-D radial grid models.

- Theoretically, finite difference models permit discretization of the reservoir using both structured and more complex unstructured grids to accurately represent the geometry of the reservoir. Local grid refinements (a finer grid embedded inside of a coarse grid) are also a feature provided by many simulators to more accurately represent the near wellbore multi-phase flow affects. This "refined meshing" near wellbores is extremely important when analysing issues such as water and gas coning in reservoirs.
- Representation of faults and their transmissibilities are advanced features provided in many simulators. In these models, inter-cell flow transmissibilities must be computed for non-adjacent layers outside of conventional neighbour-to-neighbour connections.
- Natural fracture simulation (known as dual-porosity and dual-permeability) is an advanced feature which model hydrocarbons in tight matrix blocks. Flow occurs from the tight matrix blocks to the more permeable fracture networks that surround the blocks, and to the wells.
- A black oil simulator does not consider changes in composition of the hydrocarbons as the field is produced. The compositional model, is a more complex model, where the PVT properties of oil and gas phases have been fitted to an equation of state (EOS), as a mixture of components. The simulator then uses the fitted EOS equation to dynamically track the movement of both phases and components in field.

The simulation model computes the saturation change of three phases (oil, water and gas) and pressure of each phase in each cell at each time step. As a result of declining pressure as in a reservoir depletion study, gas will be liberated from the oil. If pressures increase as a result of water or gas injection, the gas is re-dissolved into the oil phase.

A simulation project of a developed field, usually requires "history matching" where historical field production and pressures are compared to calculated values. In recent years optimisation tools such as MEPO has helped to accelerate this process, as well as improve the quality of the match obtained. The model's parameters are adjusted until a reasonable match is achieved on a field basis and usually for all wells.

Commonly, producing water cuts or water-oil ratios and gas-oil ratios are matched.

Other Engineering Approaches

Without FD models, recovery estimates and oil rates can also be calculated using numerous analytical techniques which include material balance equations (including Havlena-Odeh and Tarner method), fractional flow curve methods (1-D displacement by Buckley-Leverett, Deitz method for inclined structures, coning models), sweep efficiency estimation techniques for water floods and decline curve analysis.

These methods were developed and used prior to traditional or "conventional" simulations tools as computationally inexpensive models based on simple homogeneous reservoir description. Analytical methods generally cannot capture all the details of the given reservoir or process, but are typically numerically fast and at times, sufficiently reliable. In modern reservoir engineering, they are generally used as screening or preliminary evaluation tools. Analytical methods are especially suitable for potential assets evaluation when the data are limited and the time is critical, or for broad studies as a pre-screening tool if a large number of processes and/or technologies are to be evaluated. The analytical methods are often developed and promoted in the academia or in-house, however commercial packages also exist.

Seismic to Simulation

Seismic to Simulation is the process and associated techniques used to develop highly accurate static and dynamic 3D models of hydrocarbon reservoirs for use in predicting future production, placing additional wells, and evaluating alternative reservoir management scenarios. The process is successful if the model accurately reflects the original well logs, seismic data and production history.

Reservoir models are constructed to gain a better understanding of the subsurface that leads to informed well placement, reserves estimation and production planning. Models are based on measurements taken in the field, including well logs, seismic surveys, and production history.

Seismic to simulation enables the quantitative integration of all field data into an updateable reservoir model built by a team of geologists, geophysicists, and engineers. Key techniques used in the process include integrated petrophysics and rock physics to determine the range of lithotypes and rock properties, geostatistical inversion to determine a set of plausible seismic-derived rock property models at sufficient vertical resolution and heterogeneity for flow simulation, stratigraphic grid transfer to accurately move seismic-derived data to the geologic model, and flow simulation for model validation and ranking to determine the model that best fits all the data.

Rock Physics and Petrophysics

The first step in seismic to simulation is establishing a relationship between petrophysical key rock properties and elastic properties of the rock. This is required in order to find common ground between the well logs and seismic data.

Well logs are measured in depth and provide high resolution vertical data, but no insight into the inter-well space. Seismic are measured in time and provide great lateral detail but is quite limited in its vertical resolution. When correlated, well logs and seismic can be used to create a fine-scale 3D model of the subsurface.

Insight into the rock properties comes from a combination of basic geologic understanding and well bore measurements. Based on an understanding of how the area was formed over time, geologists can predict the types of rock likely to be present and how rapidly they vary spatially. Well log and core measurements provide samples to verify and fine-tune that understanding.

Seismic data is used by petrophysicists to identify the tops of various lithotypes and the distribution of rock properties in the inter-well space using seismic inversion attributes such as impedance. Seismic surveys measure acoustic impedance contrasts between rock layers. As different geologic structures are encountered, the sound wave reflects and refracts as a function of the impedance contrast between the layers. Acoustic impedance varies by rock type and can therefore be correlated to rock properties using rock physics relationships between the inversion attributes and petrophysical properties such as porosity, lithology, water saturation, and permeability.

Once well logs are properly conditioned and edited, a petrophysical rock model is generated that can be used to derive the effective elastic rock properties from fluid and mineral parameters as well as rock structure information. The model parameters are calibrated by comparison of the synthetic to the available elastic sonic logs. Calculations are performed following a number of rock physics algorithms including: Xu & White, Greenberg & Castagna, Gassmann, Gardner, modified upper and lower Hashin-Shtrikman, and Batzle & Wang.

When the petrophysical rock model is complete, a statistical database is created to describe the rock types and their known properties such as porosity and permeability. Lithotypes are described, along with their distinct elastic properties.

MCMC Geostatistical Inversion

In the next step of seismic to simulation, seismic inversion techniques combine well and seismic data to produce multiple equally plausible 3D models of the elastic properties of the reservoir. Seismic data is transformed to elastic property log(s) at every trace. Deterministic inversion techniques are used to provide a good overall view of the porosity over the field, and serve as a quality control check. To obtain greater detail needed for complex geology, additional stochastic inversion is then employed.

Geostatistical inversion procedures detect and delineate thin reservoirs otherwise poorly defined. Markov chain Monte Carlo (MCMC) based geostatistical inversion addresses the vertical scaling problem by creating seismic derived rock properties with vertical sampling compatible to geologic models.

All field data is incorporated into the geostatistical inversion process through the use of probability distribution functions (PDFs). Each PDF describes a particular input data in geostatistical terms using histograms and variograms, which identify the odds of a given value at a specific place and the overall expected scale and texture based on geologic insight.

Once constructed, the PDFs are combined using Bayesian inference, resulting in a posterior PDF that conforms to everything that is known about the field. A weighting system is used within the algorithm, making the process more objective. From the posterior PDF, realizations are generated using a Markov chain Monte Carlo algorithm. These realizations are statistically fair and produce models of high detail, accuracy and realism. Rock properties like porosity can be cosimulated from the elastic properties determined by the geostatistical inversion. This process is iterated until a best fit model is identified.

Inversion parameters are tuned by running the inversion many times with and without well data. Without the well data, the inversions are running in blind-well mode. These blind-well mode inversions test the reliability of the constrained inversion and remove potential biased.

This statistical approach creates multiple, equi-probable models consistent with the seismic, wells, and geology. Geostatistical inversion simultaneously inverts for impedance and discrete properties types, and other petrophysical properties such as porosity can then be jointly cosimulated.

The output volumes are at a sample rate consistent with the reservoir model because making synthetics of finely sampled models is the same as from well logs. Inversion properties are consistent with well log properties because the histograms used to generate the output rock properties from the inversion are based on well log values for those rock properties.

Uncertainty is quantified by using random seeds to generate slightly differing realizations, particularly for areas of interest. This process improves the understanding of uncertainty and risk within the model.

Stratigraphic Grid Transfer

Following geostatistical inversion and in preparation for history matching and flow simulation, the static model is re-gridded and up-scaled. The transfer simultaneously converts time to depth for the various properties and transfers them in 3D from the seismic grid to a corner-point grid. The relative locations of properties are preserved, ensuring data points in the seismic grid arrive in the correct stratigraphic layer in the corner point grid.

The static model built from seismic is typically orthogonal but flow simulators expect corner point grids. The corner point grid consists of cubes that are usually much coarser in the horizontal direction and each corner of the cube is arbitrarily defined to follow the major features in the grid. Converting directly from orthogonal to corner point can cause problems such as creating discontinuity in fluid flow.

An intermediate stratigraphic grid ensures that important structures are not misrepresented in the transfer. The stratigraphic grid has the same number of cells as the orthogonal seismic grid, but the boundaries are defined by stratigraphic surfaces and the cells follow the stratigraphic organization. This is a stratigraphic representation of the seismic data using the seismic interpretation to define the layers. The stratigraphic grid model is then mapped to the corner point grid by adjusting the zones.

Using the porosity and permeability models and a saturation height function, initial saturation models are built. If volumetric calculations identify problems in the model, changes are made in the petrophysical model without causing the model to stray from the original input data. For example, sealing faults are added for greater compartmentalization.

Model Validation and Ranking

In the last step of seismic to simulation, flow simulation continues the integration process by bringing in the production history. This provides a further validation of the static model against history. A representative set of the model realizations from the geostatistical inversion are history matched against production data.

If the properties in the model are realistic, simulated well bottom hole pressure behaviour should match historical (measured) well bottom hole pressure. Production flow rates and other engineering data should also match.

Based on the quality of the match, some models are eliminated. After the initial history match process, dynamic well parameters are adjusted as needed for each of the remaining models to improve the match. The final model represents the best match to original field measurements and production data and is then used in drilling decisions and production planning.

Petroleum Geology

Petroleum geology refers to the specific set of geological disciplines that are applied to the search for hydrocarbons (oil exploration).

Sedimentary Basin Analysis

Petroleum geology is principally concerned with the evaluation of seven key elements in sedimentary basins:

- Source
- Reservoir
- Seal
- Trap
- Timing
- Maturation
- Migration.

In general, all these elements must be assessed via a limited 'window' into the subsurface world, provided by one (or possibly more) exploration wells.

These wells present only a 1-dimensional segment through the Earth and the skill of inferring 3-dimensional characteristics from them is one of the most fundamental in petroleum geology. Recently, the availability of inexpensive, high quality 3D seismic data (from reflection seismology) and data from various electromagnetic geophysical techniques (such as Magnetotellurics) has greatly aided the accuracy of such interpretation. The following section discusses these elements in brief. For a more in-depth treatise, see the second half of this article below. Evaluation of the source uses the methods of geochemistry to quantify the nature of organic-rich rocks which contain the precursors to hydrocarbons, such that the type and quality of expelled hydrocarbon can be assessed.

The reservoir is a porous and permeable lithological unit or set of units that holds the hydrocarbon reserves. Analysis of reservoirs at the simplest level requires an assessment of their porosity (to calculate the volume of *in situ* hydrocarbons) and their permeability (to calculate how easily hydrocarbons will flow out of them). Some of the key disciplines used in reservoir analysis are the fields of structural analysis, stratigraphy, sedimentology, and reservoir engineering.

The seal, or *cap* rock, is a unit with low permeability that impedes the escape of hydrocarbons from the reservoir rock. Common seals include evaporites, chalks and shales. Analysis of seals involves assessment of their thickness and extent, such that their effectiveness can be quantified.

The trap is the stratigraphic or structural feature that ensures the juxtaposition of reservoir and seal such that hydrocarbons remain trapped in the subsurface, rather than escaping (due to their natural buoyancy) and being lost.

Analysis of maturation involves assessing the thermal history of the source rock in order to make predictions of the amount and timing of hydrocarbon generation and expulsion.

Finally, careful studies of migration reveal information on how hydrocarbons move from source to reservoir and help quantify the source (or *kitchen*) of hydrocarbons in a particular area.

Major Subdisciplines in Petroleum Geology

Several major subdisciplines exist in petroleum geology specifically to study the seven key elements.

Analysis of Source Rocks

In terms of source rock analysis, several facts need to be established. Firstly, the question of whether there actually *is* any source rock in the area must be answered. Delineation and identification of potential source rocks depends on studies of the local stratigraphy, palaeogeography and sedimentology to determine the likelihood of organic-rich sediments having been deposited in the past. If the likelihood of there being a source rock is thought to be high, the next matter to address is the state of thermal maturity of the source, and the timing of maturation. Maturation of source rocks depends strongly on temperature, such that the majority of oil generation occurs in the 60° to 120°C range.

Gas generation starts at similar temperatures, but may continue up beyond this range, perhaps as high as 200°C. In order to determine the likelihood of oil/gas generation, therefore, the thermal history of the source rock must be calculated. This is performed with a combination of geochemical analysis of the source rock (to determine the type of kerogens present and their maturation characteristics) and basin modelling methods, such as back-stripping, to model the thermal gradient in the sedimentary column.

Basin Analysis

A full scale basin analysis is usually carried out prior to defining leads and prospects for future drilling. This study tackles the petroleum system and studies source rock (presence and quality); burial history; maturation (timing and volumes); migration and focus; and potential regional seals and major reservoir units (that define carrier beds). All these elements are used to investigate where potential hydrocarbons might migrate towards. Traps and potential leads and prospects are then defined in the area that is likely to have received hydrocarbons.

Exploration Stage

Although a basin analysis is usually part of the first study a company conducts prior to moving into an area for future exploration, it is also sometimes conducted during the exploration phase. Exploration geology comprises all the activities and studies necessary for finding new hydrocarbon occurrence. Usually seismic (or 3D seismic) is shot and old exploration data (seismic lines, well logs, reports) are used to expand upon. Sometimes gravity and magnetic studies are conducted and oil seeps and spills are mapped to find potential areas for hydrocarbon occurrences. As soon as a significant hydrocarbon occurrence is found by an exploration- or wildcat-well the appraisal stage is set in.

Appraisal Stage

The Appraisal stage is used to delineate the extent of the discovery. Furthermore reservoir properties, connectivity, hydrocarbon type and gas-oil and oil-water contacts are determined to calculate potential recoverable volumes. This is usually done by drilling more appraisal wells around the initial exploration well. Furthermore some production tests may give insight in reservoir pressures and connectivity. While geochemical and petrophysical analysis gives information on the type (viscosity, chemistry, API, carbon content, etc) of the hydrocarbon and the nature of the reservoir (porosity, permeability, etc.).

Production Stage

After a hydrocarbon occurrence has been discovered and appraisal has indicated it is a commercial find the production stage is initiated. This stage focuses on extracting the hydrocarbons in a controlled way (without damaging the formation, within commercial favorable volumes, etc.). Production wells are drilled and completed in strategic positions. 3D seismic is usually available by this stage to target wells precisely for optimal recovery. Sometimes enhanced recovery (steam injection, pumps, etc.) is used to extract more hydrocarbons or to redevelop abandoned fields.

Analysis of Reservoir

The existence of a reservoir rock (typically, sandstones and fractured limestones) is determined through a combination of regional studies (i.e. analysis of other wells in the area), stratigraphy and sedimentology (to quantify the pattern and extent of sedimentation) and seismic interpretation. Once a possible hydrocarbon reservoir is identified, the key physical characteristics of a reservoir that are of interest to a hydrocarbon explorationist are its bulk rock volume, net-to-gross ratio, porosity and permeability.

Bulk rock volume, or the gross rock volume of rock above any hydrocarbon-water contact, is determined by mapping and correlating sedimentary packages. The net-to-gross ratio, typically estimated from analogues and wireline logs, is used to calculate the proportion of the sedimentary packages that contains reservoir rocks. The bulk rock volume multiplied by the net-to-gross ratio gives the net rock volume of the reservoir. The net rock volume multiplied by porosity gives the total hydrocarbon pore volume i.e. the volume within the sedimentary package that fluids (importantly, hydrocarbons and water) can occupy. The summation of these volumes for a given exploration prospect will allow explorers and commercial analysts to determine whether a prospect is financially viable.

Traditionally, porosity and permeability were determined through the study of hand specimens, contiguous parts of the reservoir that outcrop at the surface and by the technique of formation evaluation using wireline tools passed down the well itself. Modern advances in seismic data acquisition and processing have meant that seismic attributes of subsurface rocks are readily available and can be used to infer physical/sedimentary properties of the rocks themselves.

Petrophysics

Petrophysics is the study of the physical and chemical properties that describe the occurrence and behaviour of rocks, soils and fluids. Petrophysics mainly studies reservoirs of resources, including ore deposits and oil or natural gas reservoirs.

Petrophysicists in the oil and gas industry typically are employed in helping the engineers and other geoscientists understand the rock properties of the reservoir. Petrophysicists evaluate the reservoir rock properties by employing well log measurements, in which a string of measurement tools are inserted in the borehole, core measurements, in which rock samples are retrieved from subsurface, and sometimes seismic measurements, and combining them with geology and geophysics.

Petrophysical studies are used by petroleum engineering, geology, mineralogy, exploration geophysics and other related studies.

Some of the key properties studied in petrophysics are lithology, porosity, water saturation, permeability, density, solid mechanics, magnetization, electrical conductivity, thermal conductivity and radioactivity.

While most petrophysicists work in the oil and gas industry, some also work in the mining and water resource industries. The properties measured or computed fall into three broad categories: conventional petrophysical properties, rock mechanical properties, and ore quality.

Conventional Petrophysical Properties

Most petrophysicists are employed to compute what are commonly called conventional (or reservoir) petrophysical properties. These are:

Lithology: What type of rock is it. When combined with local geology and core study, geoscientists can use log measurements such as natural gamma, neutron, density, Photoelectric, resistivity or their combination to determine the lithology downhole.

Porosity: The amount of pore (or fluid occupied) space in the rock. This is typically measured using an instrument that measures the reaction of the rock to bombardment by neutrons or by gamma rays. Sonic wave speed and NMR logs are also measured to derive rock porosity.

Water saturation: The fraction of the pore space occupied by water. This is typically measured using an instrument that measures the resistivity of the rock.

Permeability: The quantity of fluid (usually hydrocarbon) that can flow from the rock as a function of time and pressure. Formation testing is so far the only direct downhole permeability measurement. In case of its absence, which is common in most cases, permeability estimation is derived from other measurements, such as porosity, NMR, sonic, by empirical correlations.

Thickness of rock with enough permeability to deliver fluids to a well bore. This property is often called "*Net reservoir rock.*" In the oil and gas industry, another quantity "*Net Pay*" is computed which is the thickness of rock that can deliver hydrocarbons to the well bore at a profitable rate.

Reservoir models are built upon their measured and derived properties to estimate the amount of hydrocarbon present in the reservoir, the rate at which that hydrocarbon can be produced to the Earth's surface through wellbores and the fluid flow in rocks. In the water resource industry, similar models are used to compute how much water can be produced to the surface over long periods of time, without depleting the aquifer.

Rock Mechanical Properties

Some petrophysicists use acoustic and density measurements of rocks to compute their mechanical properties and strength. They measure the compressional (P) wave velocity of sound through the rock and the shear (S) wave velocity and use these with the density of the rock to compute: The rocks *compressive strength* which is the compressive stress that causes a rock to fail. The rocks *flexibility*, the relationship between stress and deformation for a rock. Converted-wave analysis is also used to determine subsurface lithology and porosity. These measurements are useful to design programs to drill wells that produce oil and gas. The measurements are also used to design dams, roads, foundations for buildings and many other large construction projects. They can also be used to help interpret seismic signals from the Earth, either man-made seismic signals or those from earthquakes.

Ore Quality

Bore holes can be drilled into ore bodies (for example coal seams or gold ore) and either rock samples taken to determine the ore or coal quality at each bore hole location or the wells can be wireline logged to make measurements that can be used to infer quality. Some petrophysicists do this sort of analysis. The information is mapped and used to make mine development plans.

Methods of Analysis

Coring and core analysis is a direct measurement of petrophysical properties. In the petroleum industry rock samples are retrieved from subsurface and measured by core labs of oil company or some commercial core measurement service companies. This process is time consuming and expensive, thus can not be applied to all the wells drilled in a field.

The first "track", shows the natural gamma radiation level of the rock. The gamma radiation level "log" shows increasing radiation to the right and decreasing radiation to the left. The rocks emitting less radiation have more yellow shading. The detector is very sensitive and the amount of radiation is very low. In clastic rock formations, rocks that have smaller amounts of radiation are more likely to be coarser grained and have more pore space, rocks with higher amounts of radiation are more likely to have finer grains and less pore space.

The second track over in the plot records the depth below the reference point which is usually the Kelly bush or rotary table in feet, so these rocks are 11,900 feet below the surface of earth.

In the third track, the electrical resistivity of the rock is presented. The water in this rock is salty and the salt in the water causes the water to be electrically conductive such that lower resistivity is caused by increasing water saturation and decreasing hydrocarbon saturation.

The fourth track, shows the computed water saturation, both as "total" water (including the water bound to the rock) in magenta and the "effective water" or water that is free to flow in black. Both quantities are given as a fraction of the total pore space.

The fifth track shows the fraction of the total rock that is pore space, filled with fluids. The display of the pore space is divided into green for oil and blue for movable water. The black line shows the fraction of the pore space which contains either water or oil that can move, or be "produced." In addition to the what is included in blackline, the magenta line includes the water that is permanently bound to the rock.

The last track is a representation of the solid portion of the rock. The yellow pattern represents the fraction of the rock (excluding fluids) that is composed of coarser grained sandstone. The gray pattern represents the fraction of rock that is composed of finer grained "shale."

The sandstone is the part of the rock that contains the producible hydrocarbons and water.

Rock Volumetric Model for Shaly Sand Formation

The following definition and petrophysics model are a typical shaly sand formation model which assumes: 1. Shale is composed of silt, clay and their bounded water which will not flow. 2. Hydrocarbon are stored only in pore space in sand matrix.

ΦT - Total porosity (PHIT), which includes the pore space in sand and shale.

Sw - Total water saturation, the fraction of the pore space occupied by water.

Φe - Effective shale corrected porosity which includes only the pore space in sand. The pore space in shale which is filled with bounded water is excluded.

Swe - Effective shale corrected water saturation. The volumetric fraction of Φe which is occupied by water.

Vsh - Volumetric fraction of shale. This includes medium to very fine silt plus clay and the shale bound water.

Φsh - Shale porosity. Volumetric fraction of pore space in shale. These pore space is filled with bounded water by definition.

Key equations:

$$(1-\Phi e)(1-Vsh) + Vsh + (1-Vsh)*\Phi e*Swe + (1-Vsh)[\Phi e*(1-Swe)] = 1$$

Sandstone matrix volume + shale volume + water volume in sand + hydrocarbon volume in sand = total rock volume

$$\Phi e = \Phi T - Vsh *\Phi sh.$$

Lithology

The lithology of a rock unit is a description of its physical characteristics visible at outcrop, in hand or core samples or with low magnification microscopy, such as colour, texture, grain size, or composition.

It may be either a detailed description of these characteristics or be a summary of the gross physical character of a rock. It is the basis of subdividing rock sequences into individual lithostratigraphic units for the purposes of mapping and correlation between areas. In certain applications, such as site investigations, lithology is described using a standard terminology such as in the European geotechnical standard Eurocode 7.

Rock Type

The naming of a lithology is based on the rock type. The three major rock types are sedimentary, igneous, metamorphic. Sedimentary rocks are further classified by whether they are siliciclastic or carbonate. Siliciclastic sedimentary rocks are then subcategorized based on their grain size distribution and the relative proportions of quartz, feldspar, and lithic (rock) fragments. Carbonate rocks are classified with the Dunham or Folk classification schemes according to the constituents of the carbonate rock.

The name of an igneous rock requires information on crystal size and mineralogy. This classification can often be performed with a QAPF diagram.

Metamorphic rock naming can be based on texture, protolith, metamorphic facies, and/or the locations in which they are found. Naming based on texture and a pelite (e.g., shale, mudrock) protolith can be used to define slate and phyllite. Texture-based names are schist and gneiss.

These textures, from slate to gneiss, define a continually-increasing extent of metamorphism. Metamorphic facies are defined by the pressure-temperature fields in which particular minerals form.

Additional metamorphic rock names exist: greenstone (metamorphosed basalt and other extrusive igneous rock) is a classification based on composition and being located Precambrian terranes, while quartzite is based only on composition, as quartz is too stable and homogeneous to change phase at typical metamorphic temperatures and pressures.

Grain/Clast Size

In igneous and metamorphic rocks, grain size is a measure of the sizes of the crystals in the rock. In igneous rock, this is used to determine the rate at which the material cooled: large crystals typically indicate intrusive igneous rock, while small crystals indicate that the rock was extrusive. As metamorphic reactions progress, the grains in metamorphic rocks can often be broken down into smaller grains.

In clastic sedimentary rocks, grain size is the diameter of the grains and/or clasts that constitute the rock. These are used to determine which rock naming system to use (e.g., a conglomerate, sandstone, or mudstone one). In the case of sandstones and conglomerates, which cover a wide range of grain sizes, a word describing the grain size range is added to the rock name. Examples are "pebble conglomerate" and "fine quartz arenite".

Mineralogy

In rocks in which mineral grains are large enough to be identified using a hand lens, the visible mineralogy is included as part of the description. In the case of sequences possibly including carbonates, calcite-cemented rocks or those with possible calcite veins, it is normal to test for the presence of calcite (or other forms of calcium carbonate) using dilute hydrochloric acid by looking for effervescence.

The mineralogical composition of a rock is one of the major ways in which it is classified. In general, igneous rocks can be categorized by increasing silica content as ultramafic, mafic, intermediate, or felsic, though more mineral-specific classifications also exist. Likewise, metamorphic facies, which show the degree to which a rock has been exposed to heat and pressure and are therefore important in classifying metamorphic rocks, are determined by observing the mineral phases that are present in a sample.

Colour

The colour of a rock or its component parts is a distinctive characteristic of some rocks and is always recorded, sometimes against standard colour charts, such as that produced by the Rock-Colour Chart Committee of the Geological Society of America based on the Munsell colour system.

Fabric

The fabric of a rock describes the spatial and geometric configuration of all the elements that make it up. In sedimentary rocks the main visible fabric is normally bedding and the scale and degree of development of the bedding is normally recorded as part of the description. Metamorphic rocks (apart from those created by contact metamorphism), are characterised by well-developed planar and linear fabrics. Igneous rocks may also have fabrics due either to flow or to the settling out of particular mineral phases during crystallisation, forming cumulates.

Texture

The texture of a rock describes the relationship between the individual grains or clasts that make up the rock. Sedimentary textures include the degree of sorting, grading, shape and roundness of the clasts. Metamorphic textures include those referring to the timing of growth of large metamorphic minerals relative to a phase of deformation – before deformation porphyroclast – after deformation porphyroblast. Igneous textures include those that refer to the size of the mineral grains; vitreous - glassy, without crystals, aphanitic - grains not visible, phaneritic - grains clearly visible, and porphyritic - large grains in a finer matrix.

Small-scale Structures

Rocks often contain small-scale structures (smaller than the scale of an individual outcrop). In sedimentary rocks this may include sole markings, ripple marks, mudcracks and cross-bedding. These are recorded as they are generally characteristic of a particular depositional environment and may provide information on paleocurrent directions. In metamorphic rocks associated with the deeper levels of fault zones, small scale structures such as asymmetric boudins and microfolds are used to determine the sense of displacement across the zone. In igneous rocks, small-scale structures are mostly observed in lavas such as pahoehoe versus A, a basaltic flows, and pillows showing eruption within a body of water or beneath ice.

Surficial Lithology

Unconsolidated surficial materials may also be given a lithology. This is defined by grain size and composition, and is often attached to an interpretation of how the unit formed. Surficial lithologies can be given to lacustrine, coastal, fluvial, aeolian, glacial, and recent volcanic deposits, among others. Examples of surficial lithology classifications used by the US Geological Survey are, “Glacial Till, Loamy”, “Saline Lake Sediment”, and “Eolian Sediment, Coarse-Textured (Sand Dunes)”.

Porosity

Porosity or void fraction is a measure of the void (i.e., “empty”) spaces in a material, and is a fraction of the volume of voids over the total volume, between 0–1, or as a percentage between 0–100%. The term is used in multiple fields including pharmaceutics, ceramics, metallurgy, materials, manufacturing, earth sciences and construction.

Void Fraction in Two-phase Flow

In gas-liquid two-phase flow, the void fraction is defined as the fraction of the flow-channel volume that is occupied by the gas phase or, alternatively, as the fraction of the cross-sectional area of the channel that is occupied by the gas phase. Void fraction usually varies from location to location in the flow channel (depending on the two-phase flow pattern). It fluctuates with time and its value is usually time averaged.

In separated (i.e., non-homogeneous) flow, it is related to volumetric flow rates of the gas and the liquid phase, and to the ratio of the velocity of the two phases (called *slip ratio*).

Porosity in Earth Sciences and Construction

Used in geology, hydrogeology, soil science, and building science, the porosity of a porous medium (such as rock or sediment) describes the fraction of void space in the material, where the void may contain, for example, air or water. It is defined by the ratio:

$$\phi = \frac{V_V}{V_T}$$

where V_V is the volume of void-space (such as fluids) and V_T is the total or bulk volume of material, including the solid and void components. Both the mathematical symbols φ and n are used to denote porosity.

Porosity is a fraction between 0 and 1, typically ranging from less than 0.01 for solid granite to more than 0.5 for peat and clay. It may also be represented in percent terms by multiplying the fraction by 100.

The porosity of a rock, or sedimentary layer, is an important consideration when attempting to evaluate the potential volume of water or hydrocarbons it may contain.

Sedimentary porosity is a complicated function of many factors, including but not limited to: rate of burial, depth of burial, the nature of the connate fluids, the nature of overlying sediments (which may impede fluid expulsion). One commonly used relationship between porosity and depth is given by the Athy (1930) equation:

$$\phi(z) = \phi_0 e^{-kz}$$

where φ_0 is the surface porosity, k is the compaction coefficient (m^{-1}) and z is depth (m).

A value for porosity can alternatively be calculated from the bulk density ρ_{bulk} and particle density ρ_{particle}:

$$\phi = 1 - \frac{\rho_{\text{bulk}}}{\rho_{\text{particle}}}$$

Normal particle density is assumed to be approximately 2.65 g/cm^3, although a better estimation can be obtained by examining the lithology of the particles.

Porosity and Hydraulic Conductivity

Porosity can be proportional to hydraulic conductivity; for two similar sandy aquifers, the one with a higher porosity will typically have a higher hydraulic conductivity (more open area for the flow of water), but there are many complications to this relationship. The principal complication is that there is not a direct proportionality between porosity and hydraulic conductivity but rather an inferred proportionality.

There is a clear proportionality between pore throat radii and hydraulic conductivity. Also, there tends to be a proportionality between pore throat radii and pore volume. If the proportionality between pore throat radii and porosity exists then a proportionality between porosity and hydraulic conductivity may exist.

However, as grain size and/or sorting decreases the proportionality between pore throat radii and porosity begins to fail and therefore so does the proportionality between porosity and hydraulic conductivity. For example: clays typically have very low hydraulic conductivity (due to their small pore throat radii) but also have very high porosities (due to the structured nature of clay minerals), which means clays can hold a large volume of water per volume of bulk material, but they do not release water rapidly and therefore have low hydraulic conductivity.

Sorting and Porosity

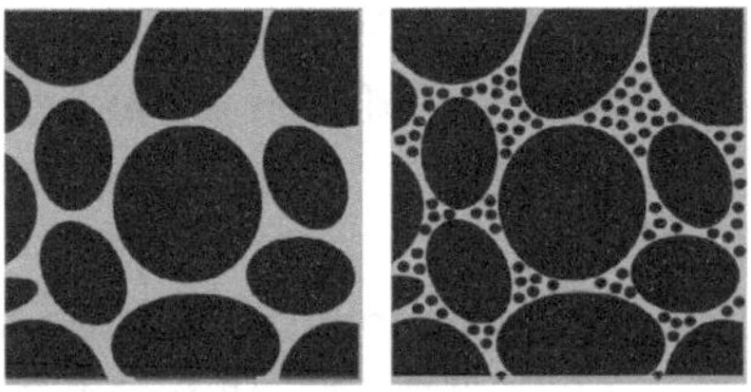

***Figure:** Effects of sorting on alluvial porosity.*

Well sorted (grains of approximately all one size) materials have higher porosity than similarly sized poorly sorted materials (where smaller particles fill the gaps between larger particles). The graphic illustrates how some smaller grains can effectively fill the pores (where all water flow takes place), drastically reducing porosity and hydraulic conductivity, while only being a small fraction of the total volume of the material.

Porosity of Rocks

Consolidated rocks (e.g. sandstone, shale, granite or limestone) potentially have more complex "dual" porosities, as compared with alluvial sediment. This can be split into connected and unconnected porosity. Connected porosity is more easily measured through the volume of gas or liquid that can flow into the rock, whereas fluids cannot access unconnected pores.

Porosity of Soil

Porosity of surface soil typically decreases as particle size increases. This is due to soil aggregate formation in finer textured surface soils when subject to soil biological processes. Aggregation involves particulate adhesion and higher resistance to compaction. Typical bulk density of sandy soil is between 1.5 and 1.7 g/cm^3.

This calculates to a porosity between 0.43 and 0.36. Typical bulk density of clay soil is between 1.1 and 1.3 g/cm^3. This calculates to a porosity between 0.58 and 0.51. This seems counterintuitive because clay soils are termed *heavy*, implying *lower* porosity. Heavy apparently refers to a gravitational moisture content effect in combination with terminology that harkens back to the relative force required to pull a tillage implement through the clayey soil at field moisture content as compared to sand.

Porosity of subsurface soil is lower than in surface soil due to compaction by gravity. Porosity of 0.20 is considered normal for unsorted gravel size material at depths below the biomantle. Porosity in finer material below the aggregating influence of pedogenesis can be expected to approximate this value.

Soil porosity is complex. Traditional models regard porosity as continuous. This fails to account for anomalous features and produces only approximate results.

Furthermore it cannot help model the influence of environmental factors which affect pore geometry. A number of more complex models have been proposed, including fractals, bubble theory, cracking theory, Boolean grain process, packed sphere, and numerous other models.

Types of Geologic Porosities

Primary Porosity

The main or original porosity system in a rock or unconfined alluvial deposit.

Secondary Porosity

A subsequent or separate porosity system in a rock, often enhancing overall porosity of a rock. This can be a result of chemical leeching of minerals or the generation of a fracture system. This can replace the primary porosity or coexist with it.

Fracture Porosity

This is porosity associated with a fracture system or faulting. This can create secondary porosity in rocks that otherwise would not be reservoirs for hydrocarbons due to their primary porosity being destroyed (for example due to depth of burial) or of a rock type not normally considered a reservoir (for example igneous intrusions or metasediments).

Vuggy Porosity

This is secondary porosity generated by dissolution of large features (such as macrofossils) in carbonate rocks leaving large holes, vugs, or even caves.

Effective Porosity

Refers to the fraction of the total volume in which fluid flow is effectively taking place and includes Caternary and dead-end (as these pores cannot be flushed, but they can cause fluid movement by release of pressure like gas expansion) pores and excludes closed pores (or non-connected cavities). This is very important for groundwater and petroleum flow, as well as for solute transport.

Ineffective Porosity

Refers to the fraction of the total volume in fluids or gases are present but in which fluid flow can not effectively take place and includes the closed pores. Understanding the morphology of the porosity is thus very important for groundwater and petroleum flow.

Dual Porosity

Refers to the conceptual idea that there are two overlapping reservoirs which interact. In fractured rock aquifers, the rock mass and fractures are often simulated as being two overlapping but distinct bodies.

Delayed yield, and leaky aquifer flow solutions are both mathematically similar solutions to that obtained for dual porosity; in all three cases water comes from two mathematically different reservoirs (whether or not they are physically different).

Macro Porosity

Refers to pores greater than 50 nm in diameter. Flow through macropores is described by bulk diffusion.

Meso Porosity

Refers to pores greater than 2 nm and less than 50 nm in diameter. Flow through mesopores is described by Knudsen diffusion.

Micro Porosity

Refers to pores smaller than 2 nm in diameter. Movement in micropores is by activated diffusion.

Measuring Porosity

Several methods can be employed to measure porosity:

- Direct methods (determining the bulk volume of the porous sample, and then determining the volume of the skeletal material with no pores (pore volume = total volume " material volume).
- Optical methods (e.g., determining the area of the material versus the area of the pores visible under the microscope). The "areal" and "volumetric" porosities are equal for porous media with random structure.
- Computed tomography method (using industrial CT scanning to create a 3D rendering of external and internal geometry, including voids. Then implementing a defect analysis utilizing computer software)
- Imbibition methods, i.e., immersion of the porous sample, under vacuum, in a fluid that preferentially wets the pores.
  - Water saturation method (pore volume = total volume of water " volume of water left after soaking).
- Water evaporation method (pore volume = (weight of saturated sample " weight of dried sample)/density of water)
- Mercury intrusion porosimetry (several non-mercury intrusion techniques have been developed due to toxicological concerns, and the fact that mercury tends to form amalgams with several metals and alloys).
- Gas expansion method. A sample of known bulk volume is enclosed in a container of known volume. It is connected to another container with a known volume which is evacuated (i.e., near vacuum pressure).

  When a valve connecting the two containers is opened, gas passes from the first container to the second until a uniform pressure distribution is attained. Using ideal gas law, the volume of the pores is calculated as

$$V_V = V_T - V_a - V_b \frac{P_2}{P_2 - P_1},$$

where

V_V is the effective volume of the pores,

V_T is the bulk volume of the sample,

V_a is the volume of the container containing the sample,

V_b is the volume of the evacuated container,

P_1 is the initial pressure in the initial pressure in volume V_a and V_V, and

P_2 is final pressure present in the entire system.

The porosity follows straightforwardly by its proper definition

$$\phi = \frac{V_V}{V_T}.$$

Note that this method assumes that gas communicates between the pores and the surrounding volume. In practice, this means that the pores must not be closed cavities.

- Thermoporosimetry and cryoporometry. A small crystal of a liquid melts at a lower temperature than the bulk liquid, as given by the Gibbs-Thomson equation.

  Thus if a liquid is imbibed into a porous material, and frozen, the melting temperature will provide information on the pore-size distribution.

  The detection of the melting can be done by sensing the transient heat flows during phase-changes using differential scanning calorimetry - (DSC thermoporometry), measuring the quantity of mobile liquid using nuclear magnetic resonance - (NMR cryoporometry) or measuring the amplitude of neutron scattering from the imbibed crystalline or liquid phases - (ND cryoporometry).

Water Content

Water content or moisture content is the quantity of water contained in a material, such as soil (called soil moisture), rock, ceramics, fruit, or wood. Water content is used in a wide range of scientific and technical areas, and is expressed as a ratio, which can range from 0 (completely dry) to the value of the materials' porosity at saturation. It can be given on a volumetric or mass (gravimetric) basis.

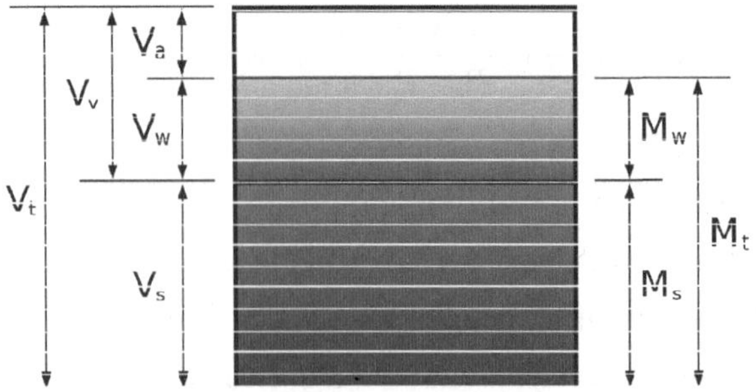

Figure 3 : Soil composition by phase: s-soil (dry), v-void (pores filled with water or air), w-water, a-air. V is volume, M is mass.

Volumetric water content, θ, is defined mathematically as:

$$\theta = \frac{V_w}{V_T}$$

where V_w is the volume of water and $V_T = V_s + V_v = V_s + V_w + V_a$ is the total volume (that is soil volume + water volume + air space).

Gravimetric water content is expressed by mass (weight) as follows:

$$u = \frac{m_w}{m_b}$$

where m_w is the mass of water and m_b is the bulk mass. The bulk mass is taken as the total mass, except for geotechnical and soil science applications where oven-dried soil he gravimetric water content by the bulk specific gravity of the material.

Other Definitions

Degree of Saturation

In soil mechanics and petroleum engineering, the term water saturation or degree of saturation, S_w is used, defined as;

$$S_w = \frac{V_w}{V_v} = \frac{V_w}{V_T \phi} = \frac{\theta}{\phi}$$

where $\varphi = V_v/V_T$ is the porosity and V_v is the volume of void or pore space.

Values of S_w can range from 0 (dry) to 1 (saturated). In reality, S_w never reaches 0 or 1 - these are idealizations for engineering use.

Normalized Volumetric Water Content

The normalized water content, Θ, (also called effective saturation or S_e) is a dimensionless value defined by van Genuchten as:

$$\Theta = \frac{\theta - \theta_r}{\theta_s - \theta_r}$$

where θ is the volumetric water content; θ_r is the residual water content, defined as the water content for which the gradient $d\theta/dh$ becomes zero; and, θ_s is the saturated water content, which is equivalent to porosity, φ.

Measurement

Direct Methods

Water content can be directly measured using a known volume of the material, and a drying oven. Volumetric water content, θ, is calculated using:

$$\theta = \frac{m_{\text{wet}} - m_{\text{dry}}}{\rho_w \cdot V_b}$$

where

m_{wet} and m_{dry} are the masses of the sample before and after drying in the oven;

ρ_w is the density of water; and

V_b is the volume of the sample before drying the sample.

For materials that change in volume with water content, such as coal, the water content, u, is expressed in terms of the mass of water per unit mass of the moist specimen:

$$u = \frac{m_{\text{wet}} - m_{\text{dry}}}{m_{\text{wet}}}$$

However, geotechnics requires the moisture content to be expressed as a percentage of the sample's dry weight i.e. % moisture content = u * 100

Where

$$u = \frac{m_{\text{wet}} - m_{\text{dry}}}{m_{\text{dry}}}$$

For wood, the convention is to report moisture content on oven-dry basis (i.e. generally drying sample in an oven set at 105 deg Celsius for 24 hours).

In wood drying, this is an important concept.

Laboratory Methods

Other methods that determine water content of a sample include chemical titrations (for example the Karl Fischer titration), determining mass loss on heating (perhaps in the presence of an inert gas), or after freeze drying.

In the food industry the Dean-Stark method is also commonly used. From the Annual Book of ASTM (American Society for Testing and Materials) Standards, the total evaporable moisture content in Aggregate (C 566) can be calculated with the formula:

$$p = \frac{W - D}{D}$$

where p is the fraction of total evaporable moisture content of sample, W is the mass of the original sample, and D is mass of dried sample.

Geophysical Methods

There are several geophysical methods available that can approximate *in situ* soil water content. These methods include: time-domain reflectometry (TDR), neutron probe, frequency domain sensor, capacitance probe, electrical resistivity tomography, ground penetrating radar (GPR), and others that are sensitive to the physical properties of water. Geophysical sensors are often used to monitor soil moisture continuously in agricultural and scientific applications.

Satellite Remote Sensing Method

Satellite microwave remote sensing is used to estimate soil moisture based on the large contrast between the dielectric properties of wet and dry soil. The microwave radiation is not sensitive to atmospheric variables, and can penetrate through clouds.

Also, microwave signal can penetrate, to a certain extent, the vegetation canopy and retrieve information from ground surface. The data from microwave remote sensing satellite such as: WindSat, AMSR-E, RADARSAT, ERS-1-2, Metop/ASCAT are used to estimate surface soil moisture.

Classification and Uses

Moisture may be present as adsorbed moisture at internal surfaces and as capillary condensed water in small pores. At low relative humidities, moisture consists mainly of adsorbed water. At higher relative humidities, liquid water becomes more and more important, depending on the pore size. In wood-based materials, however, almost all water is adsorbed at humidities below 98% RH.

In biological applications there can also be a distinction between physisorbed water and "free" water — the physisorbed water being that closely associated with and relatively difficult to remove from a biological material. The method used to determine water content may affect whether water present in this form is accounted for. For a better indication of "free" and "bound" water, the water activity of a material should be considered.

Water molecules may also be present in materials closely associated with individual molecules, as "water of crystallization", or as water molecules which are static components of protein structure.

Earth and Agricultural Sciences

In soil science, hydrology and agricultural sciences, water content has an important role for groundwater recharge, agriculture, and soil chemistry. Many recent scientific research efforts have aimed toward a predictive-understanding of water content over space and time.

Observations have revealed generally that spatial variance in water content tends to increase as overall wetness increases in semiarid regions, to decrease as overall wetness increases in humid regions, and to peak under intermediate wetness conditions in temperature regions.

There are four standard water contents that are routinely measured and used, which are described in the following table:

Name	*Notation*	*Suction pressure (J/kg or kPa)*	*Typical water content(vol/vol)*	*Description*
Saturated water content	θ_s	0	0.2–0.5	Fully saturated soil, equivalent to effective porosity
Field capacity	θ_{fc}	–33	0.1–0.35	Soil moisture 2–3 days after a rain or irrigation
Permanent wilting point	θ_{pwp} or θ_{wp}	–1500	0.01–0.25	Minimum soil moisture at which a plant wilts
Residual water content	θ_r	$-\infty$	0.001–0.1	Remaining water at high tension

And lastly the available water content, θ_a, which is equivalent to:

$$\theta_a \equiv \theta_{fc} - \theta_{pwp}$$

which can range between 0.1 in gravel and 0.3 in peat.

Agriculture

When a soil gets too dry, plant transpiration drops because the water is becoming increasingly bound to the soil particles by suction. Below the wilting point plants are no longer able to extract water. At this point they wilt and cease transpiring altogether.

Conditions where soil is too dry to maintain reliable plant growth is referred to as agricultural drought, and is a particular focus of irrigation management. Such conditions are common in arid and semi-arid environments.

Some agriculture professionals are beginning to use environmental measurements such as soil moisture to schedule irrigation. This method is referred to as *smart irrigation* or *soil cultivation.*

Groundwater

In saturated groundwater aquifers, all available pore spaces are filled with water (volumetric water content = porosity). Above a capillary fringe, pore spaces have air in them too.

Most soils have a water content less than porosity, which is the definition of unsaturated conditions, and they make up the subject of vadose zone hydrogeology. The capillary fringe of the water table is the dividing line between saturated and unsaturated conditions. Water content in the capillary fringe decreases with increasing distance above the phreatic surface.

One of the main complications which arises in studying the vadose zone, is the fact that the unsaturated hydraulic conductivity is a function of the water content of the material. As a material dries out, the connected wet pathways through the media become smaller, the hydraulic conductivity decreasing with lower water content in a very non-linear fashion.

A water retention curve is the relationship between volumetric water content and the water potential of the porous medium. It is characteristic for different types of porous medium. Due to hysteresis, different wetting and drying curves may be distinguished.

Density

The mass density or density of a material is defined as its mass per unit volume. The symbol most often used for density is ρ (the Greek letter rho). In some cases (for instance, in the United States oil and gas industry), density is also defined as its weight per unit volume; although, this quantity is more properly called specific weight. Different materials usually have different densities, so density is an important concept regarding buoyancy, purity and packaging. Osmium and iridium are the densest known metal elements at standard conditions for temperature and pressure but not the densest materials.

Less dense fluids float on more dense fluids if they do not mix. This concept can be extended, with some care, to less dense solids floating on more dense fluids. If the average density (including any air below the waterline) of an object is less than water (1000 kg/m^3) it will float in water and if it is more than water's it will sink in water.

In some cases density is expressed as the dimensionless quantities specific gravity (SG) or relative density (RD), in which case it is expressed in multiples of the density of some other standard material, usually water or air/gas. (For example, a specific gravity less than one means that the substance floats in water.)

The mass density of a material varies with temperature and pressure. (The variance is typically small for solids and liquids and much greater for gasses.) Increasing the pressure on an object decreases the volume of the object and therefore increase its density. Increasing the temperature of a substance (with some exceptions) decreases its density by increasing the volume of that substance.

In most materials, heating the bottom of a fluid results in convection of the heat from bottom to top of the fluid due to the decrease of the density of the heated fluid. This causes it to rise relative to more dense unheated material.

The reciprocal of the density of a substance is called its specific volume, a representation commonly used in thermodynamics. Density is an intensive property in that increasing the amount of a substance does not increase its density; rather it increases its mass.

History

In a well-known but probably apocryphal tale, Archimedes was given the task of determining whether King Hiero's goldsmith was embezzling gold during the manufacture of a golden wreath dedicated to the gods and replacing it with another, cheaper alloy. Archimedes knew that the irregularly shaped wreath could be crushed into a cube whose volume could be calculated easily and compared with the mass; but the king did not approve of this.

Baffled, Archimedes took a relaxing immersion bath and observed from the rise of the water upon entering that he could calculate the volume of the gold wreath through the displacement of the water.

Upon this discovery, he leaped from his bath and went running naked through the streets shouting, "Eureka! Eureka!" (Greek "I found it"). As a result, the term "eureka" entered common parlance and is used today to indicate a moment of enlightenment.

The story first appeared in written form in Vitruvius' books of architecture, two centuries after it supposedly took place. Some scholars have doubted the accuracy of this tale, saying among other things that the method would have required precise measurements that would have been difficult to make at the time.

Mathematically, density is defined as mass divided by volume:

$$\rho = \frac{m}{V},$$

where ρ is the density, m is the mass, and V is the volume. From this equation, mass density must have units of a unit of mass per unit of volume. As there are many units of mass and volume covering many different magnitudes there are a large number of units for mass density in use.

The SI unit of kilogram per cubic metre (kg/m^3) and the cgs unit of gram per cubic centimetre (g/cm^3) are probably the most common used units for density. (The cubic centimeter can be alternately called a *millilitre* or a *cc*.) 1000kg/m^3 equals one g/cm^3. In industry, other larger or smaller units of mass and or volume are often more practical and US customary units may be used.

Further, density may be expressed in terms of weight density (the weight of the material per unit volume) or as a ratio of the density with the density of a common material such as air or water.

Measurement of Density

The density at any point of a homogeneous object equals its total mass divided by its total volume. The mass is normally measured with an appropriate scale or balance; the volume may be measured directly (from the geometry of the object) or by the displacement of a fluid. Hydrostatic weighing, for instance uses, the displacement of water due to a submerged object to determine the density of the object.

If the body is not homogeneous, then the density is a function of the position. In that case the density around any given location is determined by calculating the density of a small volume around that location. In the limit of an infinitesimal volume the density of an inhomogeneous object at a point becomes: $\rho(r)=dm/dV$, where dV is an elementary volume at position r. The mass of the body then can be expressed as

$$m = \int_V \rho(\mathbf{r})dV.$$

The density of granular material can be ambiguous, depending on exactly how its volume is defined, and this may cause confusion in measurement. A common example is sand: if it is gently poured into a container, the density will be low; if the same sand is then compacted, it will occupy less volume and consequently exhibit a greater density. This is because sand, like all powders and granular solids, contains a lot of air space in between individual grains.

The density of the material including the air spaces is the bulk density, which differs significantly from the density of an individual grain of sand with no air included.

Changes of Density

In general, density can be changed by changing either the pressure or the temperature. Increasing the pressure always increases the density of a material. Increasing the temperature generally decreases the density, but there are notable exceptions to this generalization. For example, the density of water increases between its melting point at 0 °C and 4 °C; similar behaviour is observed in silicon at low temperatures.

The effect of pressure and temperature on the densities of liquids and solids is small. The compressibility for a typical liquid or solid is 10^{-6} bar^{-1} (1 bar=0.1 MPa) and a typical thermal expansivity is 10^{-5} K^{-1}. This roughly translates into needing around ten thousand times atmospheric pressure to reduce the volume of a substance by one percent. (Although the pressures needed may be around a thousand times smaller for sandy soil and some clays.) A one percent expansion of volume typically requires a temperature increase on the order of thousands of degrees Celsius.

In contrast, the density of gases is strongly affected by pressure. The density of an ideal gas is

$$\rho = \frac{MP}{RT},$$

where M is the molar mass, P is the pressure, R is the universal gas constant, and T is the absolute temperature. This means that the density of an ideal gas can be doubled by doubling the pressure, or by halving the absolute temperature.

Table : Density of Water (at 1 atm)

Temp (°C)	Density (kg/m³)
100	958.4
80	971.8
60	983.2
40	992.2
30	995.6502
25	997.0479
22	997.7735
20	998.2071
15	999.1026
10	999.7026
4	999.9720
0	999.8395
–10	998.117
–20	993.547
–30	983.854

The density of water in kilograms per cubic metre (SI unit) at various temperatures in degrees Celsius. The values below 0 °C refer to supercooled water.

Table : Density of Air (at 1 atm)

T in °C	ρ in kg/m³
–25	1.423
–20	1.395
–15	1.368
–10	1.342
–5	1.316
0	1.293
5	1.269
10	1.247
15	1.225
20	1.204

Contd...

T in °C	*ρ in kg/m³*
25	1.184
30	1.164
35	1.146

Density of Solutions

The density of a solution is the sum of mass (massic) concentrations of the components of that solution.

Mass (massic) concentration of a given component ρ_i in a solution can be called partial density of that component.

$$\rho = \sum_i \rho_i$$

Expressed as a function of the densities of pure components of the mixture and their volume participation, it reads:

$$\rho = \sum_i \rho_{i,0} \frac{V_o i}{V}.$$

Table: *Densities of Various Materials*

Material	*ρ in kg/m³*	*Notes*
Interstellar medium	$10^{-25} - 10^{-15}$	Assuming 90% H, 10% He; variable T
Earth's atmosphere	1.2	At sea level
Aerogel	1 – 2	
Styrofoam	30 – 120	
liquid hydrogen	70	
Cork	220 – 260	
Potassium	860	At STP
Sodium	970	At STP
Ice	916.7	
Water (fresh)	1000	At STP
Water (salt)	1030	
Plastics	850 – 1400	For polypropylene and PETE/PVC
Magnesium	1740	At STP
Beryllium	1850	At STP

Contd...

Material	***ρ in kg/m³***	***Notes***
Glycerol	1261	
Silicon	2330	At STP
Aluminium	2700	At STP
Diamond	3500	At STP
Titanium	4540	At STP
Selenium	4800	At STP
The Earth	5515.3	Mean density
Vanadium	6100	At STP
Antimony	6690	At STP
Zinc	7000	At STP
Chromium	7200	At STP
Manganese	7210 - 7440	At STP
Tin	7310	At STP
Iron	7870	At STP
Niobium	8570	At STP
Cadmium	8650	At STP
Cobalt	8900	At STP
Nickel	8900	At STP
Copper	8920 – 8960	Near room temperature
Bismuth	9750	At STP
Molybdenum	10220	At STP
Silver	10500	At STP
Lead	11340	Near room temperature
Thorium	11700	At STP
Rhodium	12410	At STP
The Inner Core of the Earth	~13000	As listed in Earth
Mercury	13546	At STP
Tantalum	16600	At STP
Uranium	18800	At STP
Tungsten	19300	At STP
Gold	19320	At STP
Plutonium	19840	At STP

Contd...

Material	***ρ in kg/m³***	***Notes***
Platinum	21450	At STP
Iridium	22420	At STP
Osmium	22570	At STP
The core of the Sun	~150000	
White dwarf star	1×10^9	
Atomic nuclei	2.3×10^{17}	Does not depend strongly on size of nucleus
Neutron star	$8.4 \times 10^{16} - 1 \times 10^{18}$	
Black hole	4×10^{17}	Mean density inside the Schwarzschild radius of an Earth-mass black hole (theoretical)

Density of Composite Material

In the United States, ASTM specification D792-00 describes the steps to calculate the density of a composite material.

$$\rho = \frac{W_a}{W_a + W_w - W_b}(\rho_{water})$$

where:

ρ is the density of the composite material, in g/cm³

and

W_a is the weight of the specimen when hung in the air

W_w is the weight of the partly immersed wire holding the specimen

W_b is the weight of the specimen when immersed fully in distilled water, along with the partly immersed wire holding the specimen

ρ_{water} is the density in g/cm³ of the distilled water at testing temperature (for example 0.9975 g/cm³ at 23 °C)

Other Common Units

The SI unit for density is:

- kilograms per cubic metre (kg/m³)

Litres and metric tons are not part of the SI, but are acceptable for use with it, leading to the following units:

- kilograms per litre (kg/L)
- grams per millilitre (g/mL)
- metric tons per cubic metre (t/m^3)

Densities using the following metric units all have exactly the same numerical value, one thousandth of the value in (kg/m^3). Liquid water has a density of about 1 kg/dm^3, making any of these SI units numerically convenient to use as most solids and liquids have densities between 0.1 and 20 kg/dm^3.

- kilograms per cubic decimetre (kg/dm^3)
- grams per cubic centimetre (g/cc, gm/cc or g/cm^3)
- megagrams per cubic metre (Mg/m^3)

In U.S. customary units density can be stated in:

- Avoirdupois ounces per cubic inch (oz/cu in)
- Avoirdupois pounds per cubic inch (lb/cu in)
- pounds per cubic foot (lb/cu ft)
- pounds per cubic yard (lb/cu yd)
- pounds per U.S. liquid gallon (lb/gal)
- pounds per U.S. bushel (lb/bu)
- slugs per cubic foot.

In principle there are Imperial units different from the above as the Imperial gallon and bushel differ from the U.S. units, but in practice they are no longer used, though found in older documents. The density of precious metals could conceivably be based on Troy ounces and pounds, a possible cause of confusion.

Nuclear Magnetic Resonance in Porous Media

Nuclear magnetic resonance (NMR) in porous media covers the application of using NMR as a tool to study the structure of porous media and various processes occurring in them. This technique allows the determination of characteristics such as the porosity and pore size distribution, the permeability, the water saturation, the wettability, etc. An example of this is the development of the NMR well logging device for measuring nuclear magnetic properties of fluids in underground porous sediments.

Theory of Relaxation Time Distribution in Porous Media

Microscopically the volume of a single pore in a porous media may be divided into two regions; surface area S and bulk volume V.

The surface area is a thin layer with thickness δ of a few molecules close to the pore wall surface. The bulk volume is the remaining part of the pore volume and usually dominates the overall pore volume. With respect to NMR excitations of nuclear states for hydrogen-containing molecules in these regions, different relaxation times for the induced excited energy states are expected.

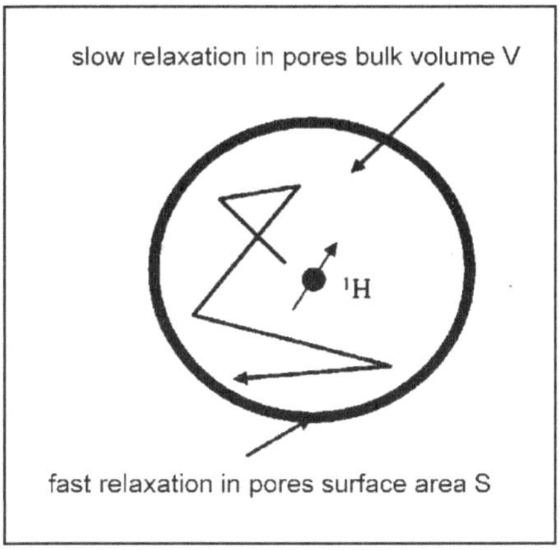

Figure: *Nuclear spin relaxation properties in a simplified pore are divided into bulk volume V and pore surface area S.*

The relaxation time is significantly shorter for a molecule in the surface area, compared to a molecule in the bulk volume. This is an effect of paramagnetic centres in the pore wall surface that causes the relaxation time to be faster. The inverse of the relaxation time T_i, is expressed by contributions from the bulk volume V, the surface area S and the self-diffusion d:

$$\frac{1}{T_i} = \left(1 - \frac{\delta S}{V}\right)\frac{1}{T_{ib}} + \frac{\delta S}{V}\frac{1}{T_{is}} + D\frac{\left(\gamma G t_E\right)^2}{12} \qquad \text{with } i = 1,2$$

where δ is the thickness of the surface area, S is the surface area, V is the pore volume, T_{ib} is the relaxation time in the bulk volume, T_{is} is the relaxation time for the surface, γ is the gyromagnetic ratio, G is the magnetic field gradient (assumed to be constant), t_E is the time between echoes and D is the self-diffusion coefficient of the fluid. The surface relaxation can be assumed as uniform or non-uniform.

The NMR signal intensity in the T_2 distribution plot reflected by the measured amplitude of the NMR signal is proportional to the total amount of hydrogen nuclei, while the relaxation time depends on the interaction between the nuclear spins and the surroundings. In a characteristic pore containing e.g. water, the bulk water exhibits a single exponential decay. The water close to the pore wall surface exhibits faster T_2 relaxation time for this characteristic pore size

NMR Permeability Correlations

NMR techniques are typically used to predict permeability for fluid typing and to obtain formation porosity, which is independent of mineralogy.

The former application uses a surface-relaxation mechanism to relate measured relaxation spectra with surface-to-volume ratios of pores, and the latter is used to estimate permeability.

The common approach is based on the model proposed by Brownstein and Tarr. They have shown that, in the fast diffusion limit, given by the expression:

$$\rho r/D$$

where ρ is the surface relaxivity of pore wall material, r is the radius of the spherical pore and D is the bulk diffusivity. The connection between NMR relaxation measurements and petrophysical parameters such as permeability stems from the strong effect that the rock surface has on promoting magnetic relaxation. For a single pore, the magnetic decay as a function of time is described by a single exponential:

$$M(t) = M_0 e^{-t/T_2}$$

where M_0 is the initial magnetization and the transverse relaxation time T_2 is given by:

$$\frac{1}{T_2} = \frac{1}{T_{2b}} + \rho\frac{S}{V}$$

S/V is the surface-to-volume ratio of the pore, T_{2b} is bulk relaxation time of the fluid that fills the pore space, and ρ is the surface relaxation strength. For small pores or large ρ, the bulk relaxation time is small and the equation can be approximated by:

$$\frac{1}{T_2} = \frac{\rho S}{V}$$

Real rocks contain an assembly of interconnected pores of different sizes. The pores are connected through small and narrow pore throats (i.e. links) that restrict interpore diffusion. If interpore diffusion is negligible, each pore can be considered to be distinct and the magnetization within individual pores decays independently of the magnetization in neighbouring pores. The decay can thus be described as:

$$M(t) = M_0 \sum_{i=1}^{n} a_i e^{-t/T_2}$$

where a_i is the volume fraction of pores of size i that decays with relaxation time T_{2i}. The multi-exponential representation corresponds to a division of the pore space into n main groups based on S/V (surface-to-volume ratio) values.

Due to the pore size variations, a non-linear optimization algorithm with multi-exponential terms is used to fit experimental data. Usually, a weighted geometric mean, T_{2lm}, of the relaxation times is used for permeability correlations:

$$T_{2lm} = \exp\left(\frac{\sum a_i \cdot \ln T_{2i}}{\sum a_i}\right) = \sqrt[\sum a_i]{\prod T_{2i}^{a_i}}$$

T_{2lm} is thus related to an average S/V or pore size. Commonly used NMR permeability correlations as proposed by Dunn *et al.* are of the form:

$$k \approx a\Phi^b (T_{2lm})^c$$

where Φ is the porosity of the rock. The exponents b and c are usually taken as four and two, respectively. Correlations of this form can be rationalized from the Kozeny–Carman equation:

$$k \approx \frac{\Phi}{\tau}\left(\frac{V}{S}\right)^2$$

by assuming that the tortuosity τ is proportional to Φ^{1-b}. However, it is well known that tortuosity is not only a function of porosity. It also depends on the formation factor $F = \tau/\Phi$. The formation factor can be obtained from resistivity logs and is usually readily available. This has given rise to permeability correlations of the form:

$$k \approx aF^b (T_{2lm})^c$$

Standard values for the exponents $b = -1$ and $c = 2$, respectively. Intuitively, correlations of this form are a better model since it incorporates tortuosity information through F. The value of the surface relaxation strength ρ affects strongly the NMR signal decay rate and hence the estimated permeability.

Surface relaxivity data are difficult to measure, and most NMR permeability correlations assume a constant ρ. However, for heterogeneous reservoir rocks with different mineralogy, ρ is certainly not constant and surface relaxivity has been reported to increase with higher fractions of microporosity. If surface relaxivity data are available it can be included in the NMR permeability correlation as

$$k \approx aF^b (\rho T_{2lm})^c$$

T_2 *Relaxation*

For fully brine saturated porous media, three different mechanisms contribute to the relaxation: bulk fluid relaxation, surface relaxation, and relaxation due to gradients in the magnetic field. In the absence of magnetic field gradients, the equations describing the relaxation are:

$$\frac{\delta M}{\delta t} = D_0 \nabla^2 M - \frac{M}{T_{2b}}$$

$$D_0 \nabla M + \rho M = 0 \text{ on S}$$

with the initial condition

$$t = 0 \text{ and } M = M_0$$

where D_0 is the self-diffusion coefficient. The governing diffusion equation can be solved by a 3D random walk algorithm. Initially, the walkers are launched at random positions in the pore space. At each time step, Δt, they advance from their current position, $x(t)$, to a new position, $x(t + \Delta t)$, by taking steps of fixed length ε in a randomly chosen direction. The time step is given by:

$$\delta t = \frac{\varepsilon^2}{6D_0}$$

The new position is given by

$$x(t + \Delta t) = x(t)\varepsilon \sin\theta \cos\Phi$$

$$y(t + \Delta t) = y(t)\varepsilon \sin\theta \cos\Phi$$

$$z(t + \Delta t) = z(t)\varepsilon \cos\theta$$

The angles $\theta(0 \leqslant \theta \leqslant \pi)$ and $\Phi(0 \leqslant \Phi \leqslant 2\pi)$ represent the randomly selected direction for each random walker in spherical coordinates. It can be noted that θ must be distributed uniformly in the range $(0, \pi)$. If a walker encounters a pore-solid interface, it is killed with a finite probability δ.

The killing probability δ is related to the surface relaxation strength by:

$$\delta = \frac{2\varepsilon\rho}{3D_0}$$

If the walker survives, it simply bounces off the interface and its position does not change. At each time step, the fraction $p(t)$ of the initial walkers that are still alive is recorded. Since the walkers move with equal probability in all directions, the above algorithm is valid as long as there is no magnetic gradient in the system.

When protons are diffusing, the sequence of spin echo amplitudes is affected by inhomogeneities in the permanent magnetic field. This results in an additional decay of the spin echo amplitudes that depends on the echo spacing $2\Delta t$.

In the simple case of a uniform spatial gradient G, the additional decay can be expressed as a multiplicative factor:

$$g(t) = e^{-\gamma^2 G^2 D_0 (\Delta\tau)^2 t}$$

where γ is the ratio of the Larmor frequency to the magnetic field intensity. The total magnetization amplitude as a function of time is then given as:

$$M(t) = M_0\left((p(t)g(t)e^{-t/T_{2b}}\right).$$

NMR as a Tool to Measure Wettability

The wettability conditions in a porous media containing two or more immiscible fluid phases determine the microscopic fluid distribution in the pore network. Nuclear magnetic resonance measurements are sensitive to wettability because of the strong effect that the solid surface has on promoting magnetic relaxation of the saturating fluid. The idea of using NMR as a tool to measure wettability was presented by Brown and Fatt in 1956.

The magnitude of this effect depends upon the wettability characteristics of the solid with respect to the liquid in contact with the surface. Their theory is based on the hypothesis that molecular movements are slower in the bulk liquid than at the solid-liquid interface. In this solid-liquid interface the diffusion coefficient is reduced, which correspond to a zone of higher viscosity. In this higher viscosity zone, the magnetically aligned protons can more easily transfer their energy to their surroundings. The magnitude of this effect depends upon the wettability characteristics of the solid with respect to the liquid in contact with the surface.

NMR Cryoporometry for Measuring Pore Size Distributions

NMR Cryoporometry (NMRC) is a recent technique for measuring total porosity and pore size distributions. It makes use of the Gibbs-

Thomson effect : small crystals of a liquid in the pores melt at a lower temperature than the bulk liquid :

The melting point depression is inversely proportional to the pore size. The technique is closely related to that of the use of gas adsorption to measure pore sizes (Kelvin equation). Both techniques are particular cases of the Gibbs Equations (Josiah Willard Gibbs): the Kelvin Equation is the constant temperature case, and the Gibbs-Thomson Equation is the constant pressure case.

To make a Cryoporometry measurement, a liquid is imbibed into the porous sample, the sample cooled until all the liquid is frozen, and then warmed slowly while measuring the quantity of the liquid that has melted.

Thus it is similar to DSC thermoporosimetry, but has higher resolution, as the signal detection does not rely on transient heat flows, and the measurement can be made arbitrarily slowly. It is suitable for measuring pore diameters in the range 2 nm–2 μm.

Nuclear Magnetic Resonance (NMR) may be used as a convenient method of measuring the quantity of liquid that has melted, as a function of temperature, making use of the fact that the T_2 relaxation time in a frozen material is usually much shorter than that in a mobile liquid. The technique was developed at the University of Kent in the UK. It is also possible to adapt the basic NMRC experiment to provide structural resolution in spatially-dependent pore size distributions, or to provide behavioural information about the confined liquid.

Electrical Resistivity and Conductivity

Electrical resistivity (also known as resistivity, specific electrical resistance, or volume resistivity) is a measure of how strongly a material opposes the flow of electric current. A low resistivity indicates a material that readily allows the movement of electric charge. The SI unit of electrical resistivity is the ohm metre (Ωm). It is commonly represented by the Greek letter ρ (rho).

Electrical conductivity or specific conductance is the reciprocal quantity, and measures a material's ability to conduct an electric current. It is commonly represented by the Greek letter σ (sigma), but κ (esp. in electrical engineering) or γ are also occasionally used. Its SI unit is siemens per metre ($S \cdot m^{-1}$) and CGSE unit is reciprocal second (s^{-1}):

$$\sigma = \frac{1}{\rho}.$$

Definitions

Scalar form

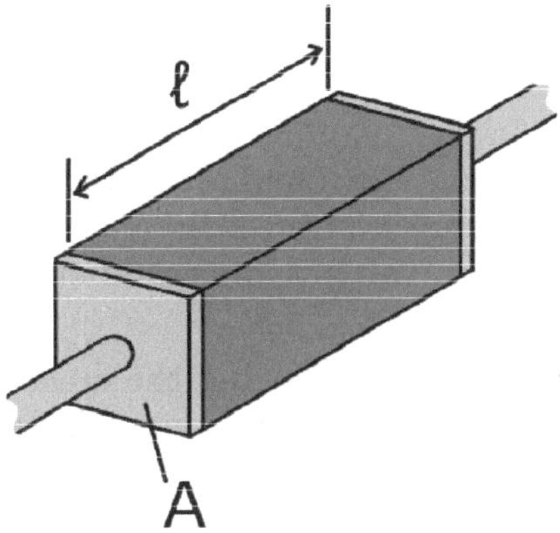

Figure: *A piece of resistive material with electrical contacts on both ends.*

Electrical resistivity ρ (Greek: rho) is defined by,

$$\rho = \frac{E}{J}$$

where

ρ is the static resistivity (measured in ohm-metres, Ω-m)

E is the magnitude of the electric field (measured in volts per metre, V/m),

J is the magnitude of the current density (measured in amperes per square metre, A/m^2),

and E and J are both inside the conductor.

Many resistors and conductors have a uniform cross section with a uniform flow of electric current and are made of one material. In this case, the above definition of ρ leads to:

$$\rho = R\frac{A}{\ell},$$

where

R is the electrical resistance of a uniform specimen of the material (measured in ohms, Ω)

l is the length of the piece of material (measured in metres, m)

ℓ is the cross-sectional area of the specimen (measured in square metres, m^2).

The reason resistivity has the dimension units of ohm-metres can be seen by transposing the definition to make resistance the subject:

$$R = \rho \frac{\ell}{A}.$$

The resistance of a given sample will increase with the length, but decrease with greater cross-sectional area. Resistance is measured in ohms. Length over area has units of 1/distance. To end up with ohms, resistivity must be in the units of "ohms × distance" (SI ohm-metre, US ohm-inch).

In a hydraulic analogy, increasing the cross-sectional area of a pipe reduces its resistance to flow, and increasing the length increases resistance to flow (and pressure drop for a given flow).

Tensor Generalization

The above scalar equation:

$$J = \sigma E$$

or equivalently

$$E = \rho J$$

is only valid for resistors and conductors which are homogeneous and isotropic, that is the conductivity is uniform throughout the conductor, carrying a uniform electric current per unit cross-sectional area J, through a uniform internal electric field E, all in one dimension only.

For all three dimensions we have at all points (r = position vector at a point in the conductor):

$$J = \rho E$$

in which (1, 2, 3 refer to components)

$$E = E(r) = \begin{pmatrix} E_1 \\ E_2 \\ E_3 \end{pmatrix} = \begin{pmatrix} E_1(r) \\ E_2(r) \\ E_3(r) \end{pmatrix}, \quad J = J(r) = \begin{pmatrix} J_1 \\ J_2 \\ J_3 \end{pmatrix} = \begin{pmatrix} J_1(r) \\ J_2(r) \\ J_3(r) \end{pmatrix}, \quad \sigma = \sigma(r).$$

This is still only true if E and J are collinear, under the same conditions as above.

Since E and J are generally vector fields, the material may not be homogeneous or isotropic so the conductivity may vary from point to point, and the E and J are not always collinear at the same point, the definition generalizes by treating conductivity as a rank-2 tensor field σ:

$$J = \sigma E,$$

where rank-2 tensors can be *represented* by matrices

$$\sigma = \sigma(r) = \begin{pmatrix} \sigma_{11} & \sigma_{12} & \sigma_{13} \\ \sigma_{21} & \sigma_{22} & \sigma_{23} \\ \sigma_{31} & \sigma_{32} & \sigma_{33} \end{pmatrix} = \begin{pmatrix} \sigma_{11}(r) & \sigma_{12}(r) & \sigma_{13}(r) \\ \sigma_{21}(r) & \sigma_{22}(r) & \sigma_{23}(r) \\ \sigma_{31}(r) & \sigma_{32}(r) & \sigma_{33}(r) \end{pmatrix},$$

equivalent to the matrix equation

$$\begin{pmatrix} J_1 \\ J_2 \\ J_3 \end{pmatrix} = \begin{pmatrix} \sigma_{11} & \sigma_{12} & \sigma_{13} \\ \sigma_{21} & \sigma_{22} & \sigma_{23} \\ \sigma_{31} & \sigma_{32} & \sigma_{33} \end{pmatrix} \begin{pmatrix} E_1 \\ E_2 \\ E_3 \end{pmatrix},$$

which allows a simple relation between components to be read off, using the Einstein summation convention;

$$J_i = \sigma_{ij} E_j.$$

The above tensor equation serves as a definition for the conductivity tensor (or field). A similar analysis for resistivity can be done, introducing a resistivity tensor (or field) in an identical form,

$$E = \rho J;$$

in which

$$\rho = \rho(r) = \begin{pmatrix} \rho_{11} & \rho_{12} & \rho_{13} \\ \rho_{21} & \rho_{22} & \rho_{23} \\ \rho_{31} & \rho_{32} & \rho_{33} \end{pmatrix} = \begin{pmatrix} \rho_{11}(r) & \rho_{12}(r) & \rho_{13}(r) \\ \rho_{21}(r) & \rho_{22}(r) & \rho_{23}(r) \\ \rho_{31}(r) & \rho_{32}(r) & \rho_{33}(r) \end{pmatrix},$$

so

$$E_i = \rho_{ij} J_j.$$

Resistivity of Various Materials

- A conductor such as a metal has high conductivity and a low resistivity.
- An insulator like glass has low conductivity and a high resistivity.
- The conductivity of a semiconductor is generally intermediate, but varies widely under different conditions, such as exposure of the material to electric fields or specific frequencies of light, and, most important, with temperature and composition of the semiconductor material.

The degree of doping in semiconductors makes a large difference in conductivity. To a point, more doping leads to higher conductivity.

The conductivity of a solution of water is highly dependent on its concentration of dissolved salts, and other chemical species that ionize in the solution.

Electrical conductivity of water samples is used as an indicator of how salt-free, ion-free, or impurity-free the sample is; the purer the water, the lower the conductivity (the higher the resistivity).

Conductivity measurements in water are often reported as *specific conductance*, relative to the conductivity of pure water at 25 °C. An EC meter is normally used to measure conductivity in a solution.

This table shows the resistivity, conductivity and temperature coefficient of various materials at 20 °C (68 °F)

Material	***ρ [Ω·m] at 20 °C***	***σ [S/m] at 20 °C***	***Temperature coefficient [K^{-1}]***
Silver	1.59×10^{-8}	6.30×10^{7}	0.0038
Copper	1.68×10^{-8}	5.96×10^{7}	0.0039
Annealed Copper[note 2]	1.72×10^{-8}	5.80×10^{7}	
Gold[note 3]	2.44×10^{-8}	4.52×10^{7}	0.0034
Aluminium[note 4]	2.82×10^{-8}	3.5×10^{7}	0.0039
Calcium	3.36×10^{-8}	2.98×10^{7}	0.0041
Tungsten	5.60×10^{-8}	1.79×10^{7}	0.0045
Zinc	5.90×10^{-8}	1.69×10^{7}	0.0037
Nickel	6.99×10^{-8}	1.43×10^{7}	0.006
Lithium	9.28×10^{-8}	1.08×10^{7}	0.006
Iron	1.0×10^{-7}	1.00×10^{7}	0.005
Platinum	1.06×10^{-7}	9.43×10^{6}	0.00392
Tin	1.09×10^{-7}	9.17×10^{6}	0.0045
Lead	2.2×10^{-7}	4.55×10^{6}	0.0039
Titanium	4.20×10^{-7}	2.38×10^{6}	X
Manganin	4.82×10^{-7}	2.07×10^{6}	0.000002
Constantan	4.9×10^{-7}	2.04×10^{6}	0.000008
Stainless steel[note 5]	6.897×10^{-7}	1.450×10^{6}	
Mercury	9.8×10^{-7}	1.02×10^{6}	0.0009
Nichrome[note 6]	1.10×10^{-6}	9.09×10^{5}	0.0004
Carbon (amorphous)	5×10^{-4} to 8×10^{-4}	1.25 to 2×10^{3}	–0.0005

Contd...

Material	***ρ [Ω·m] at 20 °C***	***σ [S/m] at 20 °C***	***Temperature coefficient [K^{-1}]***
Carbon (graphite)	$2.5e\times10^{-6}$ to 5.0×10^{-6} basal plane	2 to 3×10^{5} basal plane	
	3.0×10^{-3}//basal plane	3.3×10^{2}//basal plane	
Carbon (diamond)[note 8]	1×10^{12}	$\sim10^{-13}$	
Germanium[note 8]	4.6×10^{-1}	2.17	–0.048
Sea water[note 9]	2×10^{-1}	4.8	
Drinking water[note 10]	2×10^{1} to 2×10^{3}	5×10^{-4} to 5×10^{-2}	
Deionized water[note 11]	1.8×10^{5}	5.5×10^{-6}	
Silicon[note 8]	6.40×10^{2}	1.56×10^{-3}	–0.075
GaAs	5×10^{-7} to 10×10^{-3}	5×10^{-8} to 10^{3}	
Glass	10×10^{10} to 10×10^{14}	10^{-11} to 10^{-15}	?
Hard rubber	1×10^{13}	10^{-14}	?
Sulfur	1×10^{15}	10^{-16}	?
Air	1.3×10^{16} to 3.3×10^{16}	3 to 8×10^{-15}	
Paraffin	1×10^{17}	10^{-18}	?
Quartz (fused)	7.5×10^{17}	1.3×10^{-18}	?
PET	10×10^{20}	10^{-21}	?
Teflon	10×10^{22} to 10×10^{24}	10^{-25} to 10^{-23}	?

The effective temperature coefficient varies with temperature and purity level of the material. The 20 °C value is only an approximation when used at other temperatures. For example, the coefficient becomes lower at higher temperatures for copper, and the value 0.00427 is commonly specified at 0 °C.

The extremely low resistivity (high conductivity) of silver is characteristic of metals. George Gamow tidily summed up the nature of the metals' dealings with electrons in his science-popularizing book, *One, Two, Three...Infinity* (1947): "The metallic substances differ from all other materials by the fact that the outer shells of their atoms are bound rather loosely, and often let one of their electrons go free. Thus the interior of a metal is filled up with a large number of unattached electrons that travel aimlessly around like a crowd of displaced persons.

When a metal wire is subjected to electric force applied on its opposite ends, these free electrons rush in the direction of the force, thus forming what we call an electric current." More technically, the free electron model gives a basic description of electron flow in metals.

Temperature Dependence

In general, electrical resistivity of metals increases with temperature, while the resistivity of intrinsic semiconductors decreases with increasing temperature. In both cases, electron–phonon interactions can play a key role. At high temperatures, the resistance of a metal increases linearly with temperature. As the temperature of a metal is reduced, the temperature dependence of resistivity follows a power law function of temperature. Mathematically the temperature dependence of the resistivity ρ of a metal is given by the Bloch–Grüneisen formula:

$$\rho(T) = \rho(0) + A\left(\frac{T}{\Theta_R}\right)^n \int_0^{\frac{\Theta_R}{T}} \frac{x^n}{(e^x - 1)(1 - e^{-x})} dx$$

where ρ(0) is the residual resistivity due to defect scattering, A is a constant that depends on the velocity of electrons at the Fermi surface, the Debye radius and the number density of electrons in the metal. Θ_R is the Debye temperature as obtained from resistivity measurements and matches very closely with the values of Debye temperature obtained from specific heat measurements. n is an integer that depends upon the nature of interaction:

1. n=5 implies that the resistance is due to scattering of electrons by phonons (as it is for simple metals)
2. n=3 implies that the resistance is due to s-d electron scattering (as is the case for transition metals)
3. n=2 implies that the resistance is due to electron–electron interaction.

As the temperature of the metal is sufficiently reduced (so as to 'freeze' all the phonons), the resistivity usually reaches a constant value, known as the residual resistivity. This value depends not only on the type of metal, but on its purity and thermal history. The value of the residual resistivity of a metal is decided by its impurity concentration. Some materials lose all electrical resistivity at sufficiently low temperatures, due to an effect known as superconductivity.

An even better approximation of the temperature dependence of the resistivity of a semiconductor is given by the Steinhart–Hart equation:

$$1/T = A + B\ln(\rho) + C(\ln(\rho))^3$$

where *A*, *B* and *C* are the so-called Steinhart–Hart coefficients.

This equation is used to calibrate thermistors.

In non-crystalline semi-conductors, conduction can occur by charges quantum tunnelling from one localised site to another. This is known as variable range hopping and has the characteristic form of $\rho = Ae^{T^{-1/n}}$, where n=2,3,4 depending on the dimensionality of the system.

Complex Resistivity and Conductivity

When analyzing the response of materials to alternating electric fields, in applications such as electrical impedance tomography, it is necessary to replace resistivity with a complex quantity called impeditivity (in analogy to electrical impedance). Impeditivity is the sum of a real component, the resistivity, and an imaginary component, the reactivity (in analogy to reactance).

The magnitude of Impeditivity is the square root of sum of squares of magnitudes of resistivity and reactivity. Conversely, in such cases the conductivity must be expressed as a complex number (or even as a matrix of complex numbers, in the case of anisotropic materials) called the *admittivity*. Admittivity is the sum of a real component called the conductivity and an imaginary component called the susceptivity.

An alternative description of the response to alternating currents uses a real (but frequency-dependent) conductivity, along with a real permittivity. The larger the conductivity is, the more quickly the alternating-current signal is absorbed by the material (i.e., the more opaque the material is).

Resistivity Density Products

In some applications where the weight of an item is very important resistivity density products are more important than absolute low resistivity- it is often possible to make the conductor thicker to make up for a higher resistivity; and then a low resistivity density product material (or equivalently a high conductance to density ratio) is desirable. For example, for long distance overhead power lines—aluminium is frequently used rather than copper because it is lighter for the same conductance.

Material	*Resistivity [nΩ·m]*	*Density [g/cm³]*	*Resistivity-density product [nΩ·m·g/cm³]*
Sodium	47.7	0.97	46
Lithium	92.8	0.53	49
Calcium	33.6	1.55	52
Potassium	72.0	0.89	64

Contd...

Material	*Resistivity [nΩ·m]*	*Density [g/cm³]*	*Resistivity-density product [nΩ·m·g/cm³]*
Beryllium	35.6	1.85	66
Aluminium	26.50	2.70	72
Magnesium	43.90	1.74	76.3
Copper	16.78	8.96	150
Silver	15.87	10.49	166
Gold	22.14	19.30	427
Iron	96.1	7.874	757

Silver, although it is the least resistive metal known, has a high density and does poorly by this measure. Calcium and the alkali metals make for the best products, but are rarely used for conductors due to their high reactivity with water and oxygen. Aluminium is far more stable. And the most important attribute, the current price, excludes the best choice: Beryllium.

Thermal Conductivity

In physics, thermal conductivity, k, is the property of a material's ability to conduct heat. It appears primarily in Fourier's Law for heat conduction.

Heat transfer across materials of high thermal conductivity occurs at a faster rate than across materials of low thermal conductivity. Correspondingly materials of high thermal conductivity are widely used in heat sink applications and materials of low thermal conductivity are used as thermal insulation.

Thermal conductivity of materials is temperature dependent. In general, materials become more conductive to heat as the average temperature increases.

The reciprocal of thermal conductivity is thermal resistivity.

Units of Thermal Conductivity

In the International System of Units (SI), thermal conductivity is measured in watts per meter kelvin (W/(m ·K)).

In the imperial system of measurement thermal conductivity is measured in Btu/(hr ·ft-F) where 1 Btu/(hr ·ft-F) = 1.730735 W/(m ·K).

Other units which are closely related to the thermal conductivity are in common use in the construction and textile industries. The construction industry makes use of units such as the R-Value (resistance value) and the U-Value (thermal transmittance). Although related to the thermal conductivity of a product R and U-values are dependent on the thickness of a product.

Likewise the textile industry has several units including the Tog and the Clo which express thermal resistance of a material in a way analogous to the R-values used in the construction industry.

Note: R-Values and U-Values quoted in the US (based on the imperial units of measurement) do not correspond with and are not compatible with those used in Europe (based on the SI units of measurement).

Measurement

There are a number of ways to measure thermal conductivity. Each of these is suitable for a limited range of materials, depending on the thermal properties and the medium temperature. There is a distinction between steady-state and transient techniques. In general, steady-state techniques are useful when the temperature of the material does not change with time. This makes the signal analysis straightforward (steady state implies constant signals). The disadvantage is that a well-engineered experimental setup is usually needed. The Divided Bar (various types) is the most common device used for consolidated rock samples.

The transient techniques perform a measurement during the process of heating up. Their advantage is quicker measurements. Transient methods are usually carried out by needle probes. A method described by Angstrom involves rapidly cycling the temperature from hot to cold and back and measuring the temperature change as the heat propagates along a thin strip of material in a vacuum.

Definitions

The reciprocal of thermal conductivity is *thermal resistivity*, usually measured in kelvin-meters per watt ($K \cdot m \cdot W^{-1}$). When dealing with a known amount of material, its *thermal conductance* and the reciprocal property, *thermal resistance*, can be described. Unfortunately, there are differing definitions for these terms.

Conductance

For general scientific use, *thermal conductance* is the quantity of heat that passes in unit time through a plate of *particular area and thickness* when its opposite faces differ in temperature by one kelvin. For a plate of thermal conductivity k, area A and thickness L this is kA/L, measured in $W \cdot K^{-1}$ (equivalent to: W/°C). Thermal conductivity and conductance are analogous to electrical conductivity ($A \cdot m^{-1} \cdot V^{-1}$) and electrical conductance ($A \cdot V^{-1}$).

There is also a measure known as heat transfer coefficient: the quantity of heat that passes in unit time through *unit area* of a plate of particular thickness when its opposite faces differ in temperature by one kelvin. The reciprocal is *thermal insulance*. In summary:

- *thermal conductance* = kA/L, measured in $W \cdot K^{-1}$
  — *thermal resistance* = $L/(kA)$, measured in $K \cdot W^{-1}$ (equivalent to: °C/W)
- *heat transfer coefficient* = k/L, measured in $W \cdot K^{-1} \cdot m^{-2}$
  — *thermal insulance* = L/k, measured in $K \cdot m^2 \cdot W^{-1}$.

The heat transfer coefficient is also known as *thermal admittance.*

Resistance

When thermal resistances occur in series, they are additive. So when heat flows through two components each with a resistance of 1 °C/W, the total resistance is 2 °C/W.

A common engineering design problem involves the selection of an appropriate sized heat sink for a given heat source. Working in units of thermal resistance greatly simplifies the design calculation. The following formula can be used to estimate the performance:

$$R_{hs} = \frac{\Delta T}{P_{th}} - R_s$$

where:

- R_{hs} is the maximum thermal resistance of the heat sink to ambient, in °C/W (equivalent to K/W)
- ΔT is the temperature difference (temperature drop), in °C
- P_{th} is the thermal power (heat flow), in watts
- R_s is the thermal resistance of the heat source, in °C/W

For example, if a component produces 100 W of heat, and has a thermal resistance of 0.5 °C/W, what is the maximum thermal resistance of the heat sink? Suppose the maximum temperature is 125 °C, and the ambient temperature is 25 °C; then the ΔT is 100 °C. The heat sink's thermal resistance to ambient must then be 0.5 °C/W or less.

Transmittance

A third term, *thermal transmittance,* incorporates the thermal conductance of a structure along with heat transfer due to convection and radiation. It is measured in the same units as thermal conductance and is sometimes known as the *composite thermal conductance*. The term *U-value* is another synonym.

Influencing Factors

Temperature

The effect of temperature on thermal conductivity is different for metals and nonmetals. In metals conductivity is primarily due to lattice vibrations and free electron, however, free electrons play a dominant role. Therefore any increase in temperature increases the lattice vibrations but affects the movement of free electrons adversly thereby decreasing the conductivity.

On the other hand conductivity in nonmetals is only due to lattice vibrations which increases with increasing temperature, and so the conductivity of nonmetals increases with increasing temperature.

Material Phase

When a material undergoes a phase change from solid to liquid or from liquid to gas the thermal conductivity may change. An example of this would be the change in thermal conductivity that occurs when ice (thermal conductivity of 2.18 W/(m·K) at 0 °C) melts into liquid water (thermal conductivity of 0.58 W/(m·K) at 0 °C).

Material Structure

Pure crystalline substances exhibit very different thermal conductivities along different crystal axes, due to differences in phonon coupling along a given crystal axis.

Sapphire is a notable example of variable thermal conductivity based on orientation and temperature, with 35 W/(m·K) along the c-axis and 32 W/(m·K) along the a-axis.

Electrical Conductivity

In metals, thermal conductivity approximately tracks electrical conductivity according to the Wiedemann-Franz law, as freely moving valence electrons transfer not only electric current but also heat energy.

However, the general correlation between electrical and thermal conductance does not hold for other materials, due to the increased importance of phonon carriers for heat in non-metals.

As shown in the table below, highly electrically conductive silver is less thermally conductive than diamond, which is an electrical insulator.

Convection

Air and other gases are generally good insulators, in the absence of convection. Therefore, many insulating materials function simply by having a large number of gas-filled pockets which prevent large-scale convection. Examples of these include expanded and extruded polystyrene (popularly referred to as "styrofoam") and silica aerogel. Natural, biological insulators such as fur and feathers achieve similar effects by dramatically inhibiting convection of air or water near an animal's skin.

Light gases, such as hydrogen and helium typically have high thermal conductivity. Dense gases such as xenon and dichlorodifluoromethane have low thermal conductivity. An exception, sulfur hexafluoride, a dense gas, has a relatively high thermal conductivity due to its high heat capacity. Argon, a gas denser than air, is often used in insulated glazing (double paned windows) to improve their insulation characteristics.

Experimental Values

Thermal conductivity is important in building insulation and related fields. However, materials used in such trades are rarely subjected to chemical purity standards. Several construction materials' *k* values are listed below. These should be considered approximate due to the uncertainties related to material definitions. In the opposite end of the spectrum, solutions for computer cooling usually use high thermal capacity materials such as silver, copper and aluminium, to cool down specific components.

The following table is meant as a small sample of data to illustrate the thermal conductivity of various types of substances.

This is a list of approximate values of thermal conductivity, *k*, for some common materials. Please consult the list of thermal conductivities for more accurate values, references and detailed information.

Material	*Thermal conductivity [W/(m·K)]*
Silica Aerogel	0.004 - 0.04
Air	0.025
Wood	0.04 - 0.4
Hollow Fill Fibre Insulation	0.042
Alcohols and oils	0.1 - 0.21
Polypropylene	0.25

Contd...

Material	*Thermal conductivity [W/(m·K)]*
Mineral oil	0.138
Rubber	0.16
LPG	0.23 - 0.26
Cement, Portland	0.29
Epoxy (silica-filled)	0.30
Epoxy (unfilled)	0.12 - 0.177
Water (liquid)	0.6
Thermal grease	0.7 - 3
Thermal epoxy	1 - 7
Glass	1.1
Soil	1.5
Concrete, stone	1.7
Ice	2
Sandstone	2.4
Mercury	8.3
Stainless steel	12.11 ~ 45.0
Lead	35.3
Aluminium	237 (pure)120—180 (alloys)
Gold	318
Copper	401
Silver	429
Diamond	900 - 2320
Graphene	(4840±440) - (5300±480)

Physical Origins

Heat flux is exceedingly difficult to control and isolate in a laboratory setting. Thus at the atomic level, there are no simple, correct expressions for thermal conductivity.

Atomically, the thermal conductivity of a system is determined by how atoms composing the system interact. There are two different approaches for calculating the thermal conductivity of a system.

- The first approach employs the Green-Kubo relations. Although this employs analytic expressions which in principle can be solved, calculating the thermal conductivity of a dense fluid or solid using this relation requires the use of molecular dynamics computer simulation.

- The second approach is based upon the relaxation time approach. Due to the anharmonicity within the crystal potential, the phonons in the system are known to scatter. There are three main mechanisms for scattering:
  - Boundary scattering, a phonon hitting the boundary of a system;
  - Mass defect scattering, a phonon hitting an impurity within the system and scattering;
  - Phonon-phonon scattering, a phonon breaking into two lower energy phonons or a phonon colliding with another phonon and merging into one higher energy phonon.

Lattice Waves

Heat transport in both glassy and crystalline dielectric solids occurs through elastic vibrations of the lattice (phonons). This transport is limited by elastic scattering of acoustic phonons by lattice defects.

These predictions were confirmed by the experiments of Chang and Jones on commercial glasses and glass ceramics, where mean free paths were limited by "internal boundary scattering" to length scales of 10^{-2} cm to 10^{-3} cm.

The phonon mean free path has been associated directly with the effective relaxation length for processes without directional correlation. Thus, if V_g is the group velocity of a phonon wave packet, then the relaxation length is defined as:

$$l = V_g t$$

where t is the characteristic relaxation time. Since longitudinal waves have a much greater phase velocity than transverse waves, V_{long} is much greater than V_{trans}, and the relaxation length or mean free path of longitudinal phonons will be much greater. Thus, thermal conductivity will be largely determined by the speed of longitudinal phonons.

Regarding the dependence of wave velocity on wavelength or frequency (dispersion), low-frequency phonons of long wavelength will be limited in relaxation length by elastic Rayleigh scattering. This type of light scattering form small particles is proportional to the fourth power of the frequency.

For higher frequencies, the power of the frequency will decrease until at highest frequencies scattering is almost frequency independent.

Similar arguments were subsequently generalized to many glass forming substances using Brillouin scattering.

Equations

First, we define heat conduction, H:

$$H = \frac{\Delta Q}{\Delta t} = kA\frac{\Delta T}{x}$$

where $\frac{\Delta Q}{\Delta t}$ is the rate of heat flow, k is the thermal conductivity, A is the total cross sectional area of conducting surface, ΔT is temperature difference, and x is the thickness of conducting surface separating the two temperatures. Dimension of thermal conductivity = $M^1L^1T^{-3}K^{-1}$

Rearranging the equation gives thermal conductivity:

$$k = \frac{\Delta Q}{\Delta t}\frac{1}{A}\frac{x}{\Delta T}$$

(Note: $\Delta T/x$ is the temperature gradient)

I.E. It is defined as the quantity of heat, ΔQ, transmitted during time Δt through a thickness x, in a direction normal to a surface of area A, per unit area of A, due to a temperature difference ΔT, under steady state conditions and when the heat transfer is dependent only on the temperature gradient.

Alternatively, it can be thought of as a flux of heat (energy per unit area per unit time) divided by a temperature gradient (temperature difference per unit length)

$$k = \frac{\Delta Q}{A\Delta t}\frac{x}{\Delta T}.$$

Permeability (Earth Sciences)

Permeability in fluid mechanics and the earth sciences (commonly symbolized as κ, or k) is a measure of the ability of a porous material (often, a rock or unconsolidated material) to allow fluids to pass through it.

Units

The SI unit for permeability is m^2. A traditional unit for permeability is the *darcy* (D), or more commonly the *millidarcy* (mD) (1 darcy $\approx 10^{-12}m^2$). The unit of cm^2 is also sometimes used ($1\ m^2 = 10^4\ cm^2$).

Applications

The concept of permeability is of importance in determining the flow characteristics of hydrocarbons in oil and gas reservoirs, and of groundwater in aquifers. For a rock to be considered as an exploitable hydrocarbon reservoir without stimulation, its permeability must be greater than approximately 100 mD (depending on the nature of the hydrocarbon - gas reservoirs with lower permeabilities are still exploitable because of the lower viscosity of gas with respect to oil). Rocks with permeabilities significantly lower than 100 mD can form efficient *seals*. Unconsolidated sands may have permeabilities of over 5000 mD.

The concept has also many practical applications outside of geology, for example in chemical engineering (e.g., filtration).

Description

Permeability is part of the proportionality constant in Darcy's law which relates discharge (flow rate) and fluid physical properties (e.g. viscosity), to a pressure gradient applied to the porous media:

$$v = \frac{\kappa}{\mu}\frac{\Delta P}{\Delta x}$$

Therefore:

$$\kappa = v\frac{\mu \Delta x}{\Delta P}$$

where:

v is the superficial fluid flow velocity through the medium (i.e., the average velocity calculated as if the fluid were the only phase present in the porous medium) (m/s)

κ is the permeability of a medium (m^2)

μ is the dynamic viscosity of the fluid (Pa ·s)

ΔP is the applied pressure difference (Pa)

Δx is the thickness of the bed of the porous medium (m)

In naturally occurring materials, permeability values range over many orders of magnitude.

Relation to Hydraulic Conductivity

The proportionality constant specifically for the flow of water through a porous media is called the hydraulic conductivity; permeability is a portion of this, and is a property of the porous media only, not the fluid. Given the value of hydraulic conductivity for a subsurface system, *k*, the permeability can be calculated as:

$$\kappa = k\frac{\mu}{\rho g}$$

where

- κ is the permeability, m^2
- k is the hydraulic conductivity, m/s
- μ is the dynamic viscosity, kg/(m·s)
- ρ is the density of the fluid, kg/m^3
- g is the acceleration due to gravity, m/s^2.

Determination

Permeability is typically determined in the lab by application of Darcy's law under steady state conditions or, more generally, by application of various solutions to the diffusion equation for unsteady flow conditions.

Permeability needs to be measured, either directly (using Darcy's law) or through estimation using empirically derived formulas. However, for some simple models of porous media, permeability can be calculated (e.g., random close packing of identical spheres).

Permeability Model based on Conduit Flow

Based on Hagen–Poiseuille equation for viscous flow in a pipe, permeability can be expressed as:

$$\kappa_I = C \cdot d^2$$

where:

κ_I is the intrinsic permeability [length^2]

C is a dimensionless constant that is related to the configuration of the flow-paths

d is the average, or effective pore diameter [length].

Intrinsic and Absolute Permeability

The terms *intrinsic permeability* and *absolute permeability* states that the permeability value in question is an intensive property (not a spatial average of a heterogeneous block of material), that it is a function of the material structure only (and not of the fluid), and explicitly distinguishes the value from that of relative permeability.

Permeability to Gases

Sometimes permeability to gases can be somewhat different that those for liquids in the same media. One difference is attributable to "slippage" of gas at the interface with the solid when the gas mean free path is comparable to the pore size (about 0.01 to 0.1 μm at standard temperature and pressure).

For example, measurement of permeability through sandstones and shales yielded values from 9.0×10^{-19} m^2 to 2.4×10^{-12} m^2 for water and between 1.7×10^{-17} m^2 to 2.6×10^{-12} m^2 for nitrogen gas.

Tensor Permeability

To model permeability in anisotropic media, a permeability tensor is needed. Pressure can be applied in three directions, and for each direction, permeability can be measured (via Darcy's law in 3D) in three directions, thus leading to a 3 by 3 tensor. The tensor is realized using a 3 by 3 matrix being both symmetric and positive definite (SPD matrix):

- The tensor is symmetric by the Onsager reciprocal relations.
- The tensor is positive definite as the component of the flow parallel to the pressure drop is always in the same direction as the pressure drop.

The permeability tensor is always diagonalizable (being both symmetric and positive definite). The eigenvectors will yield the principal directions of flow, meaning the directions where flow is parallel to the pressure drop, and the eigenvalues representing the principal permeabilities.

Integrated Asset Modelling

Integrated Asset Modelling (or IAM) is the generic term used in the oil industry for computer modelling of both the subsurface and the surface elements of a field development. Historically the reservoir has always been modelled separately from the surface network and the facilities.

In order to capture the interaction between those two or more standalone models, several time consuming iterations were required. For example a change in the water breakthrough leads to a change in the deliverability of the surface network which in turn leads to a production acceleration or deceleration in the reservoir. In order to go through this lengthy process more quickly the industry has slowly been adopting a more integrated approach which captures the constraints imposed by the infrastructure on the network immediately.

Basis

As the aim of an IAM is to provide a production forecast which honours both the physical realities of the reservoir and the infrastructure it needs to contain the following elements;

- a pressure network
- a subsurface saturation model
- an availability model
- a constraint manager
- a production optimisation algorithm.

Some but not all models also contain an economics and risk module so that the IAM can be used for economic screening.

IPM vs. IAM

Historically the terms Integrated Production Modelling and Integrated Asset Modelling have been used interchangeably. The two are closely related but their focus is not the same. An IPM focuses on day to day production optimisation and is typically used for tactical decisions such as gas lift allocation. The perspective of the IAM is the whole life of field. Both need a production optimisation algorithm but for the IAM it is a means to an end, not the end itself.

Benefits of Integrated Asset Modelling

Having an IAM built of an asset or future project offers several advantages;

- Faster runtimes which allow scenario analysis and Monte Carlo analysis
- Insight in the interactions between various components of a development
- An answer in economic rather than recovery terms (not always available).

Drawbacks of Integrated Asset Modelling

By its very nature an IAM requires a multi disciplinary approach. Most companies are too compartmentalised for this to be easy, as a result of this an integrated approach has the following drawbacks;

- less accurate than the traditional approach
- difficult to spot errors
- requires constant communication between various departments, ownership is either vague or too much part of one silo.

The biggest barrier to adoption of IAM is frequently the resistance of reservoir engineers to any simplification of the subsurface.

Appropriate use of IAM

As with any other software because of the inherent limitations in any virtual model use of an IAM is only appropriate during various stages of a project life. There are no hard and fast rules for this as there are a variety of software packages on the market which offer very accurate modelling of a very small scope to very rough modelling of a very large scope and anything in between.

Currently the definition of IAM contains anything from daily optimisation to portfolio management. The success or failure of an IAM implementation project therefore depends on selecting the tool which is as complex as it needs to be but no more. The following contains some examples of areas where an IAM is the appropriate decision support tool

- Concept Select
- Debottlenecking and optimisation of very large or complex infrastructures
- Life of field analysis of production optimisation scenarios

Note that for most of these areas the accuracy of the reservoir proxy is not important, the decision is made based on relative performance differences, not absolute values.

Approach

Several different software packages are commercially available and there is a clear difference in philosophy between some of them.

Linked Existing Software

Some vendors who have previously marketed standalone software for the subsurface and the surface are now marketing additional software which provides a datalink between the various packages. The obvious benefit of this approach is that there is no loss in accuracy and it does not require a remodelling exercise. However this approach also has its drawbacks, there is no time gain and the integration component of the entire package requires expertise which is not readily available, external specialist are frequently called upon to build and maintain the links between the components.

Bespoke Software

There are relatively few software packages on the market which are truly integrated. None of these can match the accuracy of the linked existing software but they offer the benefit of shorter runtimes and lower expertise tresholds.

Software as a Service

A number of the established service companies now offer integrated asset modelling as a service. In practice this means that existing models will be either converted or linked by specialists to form an integrated solution. This solution is expensive but frequently the preferred option if the highest accuracy is required.

Core Sample

A core sample is a cylindrical section of (usually) a naturally occurring substance consistent enough to hold a layered structure. Most core samples are obtained by drilling with special drills into the substance, for example sediment or rock, with a hollow steel tube called a core drill. The hole made for the core sample is called the "core hole". A variety of core samplers exist to sample different media under different conditions. More continue to be invented on a regular basis. In the coring process, the sample is pushed more or less intact into the tube. Removed from the tube in the laboratory, it is inspected and analyzed by different techniques and equipment depending on the type of data desired.

Core samples can also be taken to test the properties of manmade materials, such as concrete, ceramics, some metals and alloys, especially the softer ones. Core samples can be taken of living things, including human beings, especially of a person's bones for microscopic examination to help diagnose diseases.

Methods

The composition of the subject materials can vary from almost liquid to the strongest materials found in nature or technology, and the location of the subject materials can vary from on the laboratory bench to over 10 km from the surface of the Earth in a borehole. The range of equipment and techniques applied to the task is correspondingly great. Core samples are most often be taken with their long axis oriented roughly parallel to the axis of a borehole, or parallel to the gravity field for the gravity-driven tools. However it is also possible to take core samples from the wall of an existing borehole. Taking samples from an exposure, albeit an overhanging rock face or on a different planet, is almost trivial. (The Mars Exploration Rovers carry a Rock Abrasion Tool which is logically equivalent to the "rotary sidewall core" tool described below.)

Some common techniques include :

- *gravity coring*, in which the core sampler is dropped into the sample, usually the bed of a water body, but essentially the same technique can also be done on soft materials on land. The penetration forces, if recorded, give information about the strength of different depths in the material, which may be the only information required, with samples as an incidental benefit. This technique is common in both civil engineering site investigations (where the techniques tend towards into pile driving) and geological studies of recent aquatic deposits. The low strength of the materials penetrated means that cores have to be relatively small.
- *vibracoring*, in which the sampler is vibrated to allow penetration into thixotropic media. Again, the physical strength of the subject material limits the size of core that can be retrieved.
- *drilling* exploration diamond drilling where a rotating annular tool backed up by a cylindrical core sample storage device is pressed against the subject materials to cut out a cylinder of the subject material. A mechanism is normally needed to retain the cylindrical sample in the coring tool. Depending on circumstances, particularly the consistency and composition of the subject materials, different arrangements may be needed within the core tools to support and protect the sample on its way to surface; it is often also necessary to control or reduce the contact between the drilling fluid and the core sample, to reduce changes from the coring process. The mechanical forces imposed on the core sample by the tool frequently lead to fracture of the core and loss of less-competent intervals, which can greatly complicate interpretation of the core. Cores can routinely be cut as small as a few millimetres in diameter (in wood, for dendrochronology) up to over 150 millimetres in diameter (routine in oil exploration). The lengths of samples can range from less than a meter (again, in wood, for dendrochronology) to around 100 metres in one run (in oil exploration), and many runs can be made in succession if "quick look" analysis in the field suggests that the zone of interest is continuing.
- *percussion sidewall coring* coring uses robust cylindrical "bullets" explosively propelled into the wall of a borehole to

retrieve a (relatively) small, short core sample. These tend to be heavily shattered, rendering porosity/permeability measurements dubious, but are often sufficient for lithological and micropalaeontological study. Many samples can be attempted in a single run of the tools, which are typically configured with 20 to 30 "bullets" and propulsive charges along the length of a tool. Several tools can often be ganged together for a single run. The success rates for firing a particular bullet, it penetrating the borehole wall, the retention system recovering the bullet from the borehole wall, and the sample being retained in the bullet are all relatively low, so it is not uncommon for only half the samples attempted to be successful. This is an important consideration in planning sample programmes.

- *rotary sidewall coring* where a miniaturised automated rotary drilling tool is applied to the side of the borehole to cut a sample similar in size to a percussion sidewall core (described above). These tend to suffer less deformation than percussion cores. However the core-cutting process takes longer and jams are common in the ancillary equipment which retrieves the sample from the drill bit and stores it within the tool body. This complicates the planning of a coring programme.

Management of Cores and Data

Although often neglected, core samples always degrade to some degree in the process of cutting the core, handling it, and studying it. Non-destructive techniques are increasingly common (e.g. the use of MRI scanning to characterise grains, pore fluids, pore spaces (porosity) and their interactions (constituting part of permeability) but such expensive subtlety is likely wasted on a core that has been shaken on an unsprung lorry for 300 km of dirt road. What happens to core between the retrieval equipment and the final laboratory (or archive) is an oft-neglected part of record keeping and core management. Equally, a core sample which cannot be related to its context (where it was before it became a core sample) has lost much of its benefit.

The identification of the borehole, and the position and orientation ("way up") of the core in the borehole is critical, even if the borehole is in a tree trunk - dendrochronologists always try to include a bark surface in their samples so that the date of most-recent growth of the tree can be unambiguously determined.

If these data become separated from core samples, it is generally impossible to regain that data. The cost of a coring operation can vary from a few currency units (for a hand-caught core from a soft soil section) to tens of millions of currency units (for sidewall cores from a remote-area offshore borehole many kilometres deep). Inadequate recording of such basic data has ruined the utility of both types of core.

Different disciplines have different local conventions of recording these data, and the user should familiarise themselves with their area's conventions. For example, in the oil industry, orientation of the core is typically recorded by marking the core with two longitudinal colour streaks, with the red one on the right when the core is being retrieved and marked at surface. Cores cut for mineral mining may have their own, different, conventions. Civil engineering or soil studies may have their own, different, conventions as their materials are often not competent enough to make permanent marks on.

It is becoming increasingly common to retain core samples in cylindrical packaging which forms part of the core-cutting equipment, and to make the marks of record on these "inner barrels" in the field prior to further processing and analysis in the laboratory. Sometimes core is shipped form the field to the laboratory in as long a length as it comes out of the ground; other times it is cut into standard lengths (5m or 1m or 3 ft) for shipping, then reassembled in the laboratory. Some of the "inner barrel" systems are capable of being reversed on the core sample, so that in the laboratory the sample goes "wrong way up" when the core is reassembled. This can complicate interpretation.

If the borehole has petrophysical measurements made of the wall rocks, and these measurements are repeated along the length of the core then the two data sets correlated, one will almost universally find that the depth "of record" for a particular piece of core differs between the two methods of measurement.

Which set of measurements to believe then becomes a matter of policy for the client (in an industrial setting) or of great controversy (in a context without an overriding authority. Recording that there are discrepancies, for whatever reason, retains the possibility of correcting an incorrect decision at a later date ; destroying the "incorrect" depth data makes it impossible to correct a mistake later. Any system for retaining and archiving data and core samples needs to be designed so that dissenting opinion like this can be retained.

If core samples from a campaign are competent, it is common practice to "slab" them - cut the sample into two or more samples longitudinally - quite early in laboratory processing so that one set of samples can be archived early in the analysis sequence as a protection against errors in processing. "Slabbing" the core into a 2/3 and a 1/3 set is common. It is also common for one set to be retained by the main customer while the second set goes to the government (who often impose a condition for such donation as a condition of exploration/exploitation licensing). "Slabbing" also has the benefit of preparing a flat, smooth surface for examination and testing of profile permeability, which is very much easier to work with than the typically rough, curved surface of core samples when they're fresh from the coring equipment. Photography of raw and "slabbed" core surfaces is routine, often under both natural and ultra-violet light.

History of Coring

The technique of coring long predates attempts to drill into the Earth's mantle by the Deep Sea Drilling Program. The value to oceanic and other geologic history of obtaining cores over a wide area of sea floors soon became apparent. Core sampling by many scientific and exploratory organizations expanded rapidly. To date hundreds of thousands of core samples have been collected from floors of all the planet's oceans and many of its inland waters.

Access to many of these samples is facilitated by the Index to Marine & Lacustrine Geological Samples,

> *"A collaboration between twenty institutions and agencies that operate geological repositories."*

The above agency keeps a record of the samples held in the repositories of its member organizations. Data includes

> *"Lithography, texture, age, principal investigator, province, weathering/metamorphism, glass remarks and descriptive comments"*

Layering

Any natural medium at or under the Earth's surface or other body that is consistent enough to maintain a solid or semi-solid structure is layered. The layering comes from successive deposition or growth in time of structural or compositional variants of the medium.

Most familiar to us are the stratigraphic layers of the Earth's surface on which the geologic history of the surface is based; for example, the Eocene, Oligocene, Miocene, etc. Each layer in this case contains distinctive fossils generated by the evolution of species. Layers often are divided into sublayers.

Layering is more pervasive than the broad outline of the Geologic Time Scale leads us to believe. Any change in environment causes a new layer to be deposited. A succession of plant species in a region, for example, causes a succession of layers containing different pollen in ice and mud. Variation in rainfall causes tree rings to be of different widths.

Informational Value of Core Samples

Scientific coring began as a method of sampling the ocean floor. It soon expanded to lakes, ice, mud, soil and wood. Cores on very old trees give information about their growth rings without destroying the tree.

Cores indicate variations of climate, species and sedimentary composition during geologic history. The dynamic phenomena of the Earth's surface are for the most part cyclical in a number of ways, especially temperature and rainfall.

There are many ways to date a core. Once dated, it gives valuable information about changes of climate and terrain. For example, cores in the ocean floor, soil and ice have altered the view of the geologic history of the Pleistocene entirely.

Bibliography

Armistead, George: *Safety in Petroleum Refining and Related Industries*, New York: John G. Simmons & Co., 1950.

Bhattacharya, K. G. and A. K. Talukdar: *Catalysis in Petroleum and Petrochemical Industries*, Alpha Science International, Delhi, 2005.

Black, Brian: *Petrolia: The Landscape of America's First Oil Boom*. Baltimore: Johns Hopkins University Press, 2000.

Bromley, Simon: *American Hegemony and World Oil: The Industry, the State System, and the World Economy*. University Park, Pa., 1991.

Chester, Edward W.: *United States Oil Policy and Diplomacy: A Twentieth-Century Overview*. West-port, Conn.: Greenwood Press, 1983.

Chisholm, Archibald H. T.: *The First Kuwait Oil Concession Agreement: A Record of the Negotiations,* 1911-1934. London: Frank Cass, 1975.

Davenport, E. H. and Sidney Russell Cooke: *The Oil Trusts and Anglo-American Relations*. Macmillan, 1924

Diwan, Parag and Ashutosh Karnatak: *Handbook of Natural Gas Technology and Business*, Pentagon Energy Press, Delhi, 2009.

Eaton, Rev. S.J.M.: *Petroleum, a History of the Oil Region*, J.P. Skelly & Co., Philadelphia, 1866.

Engler, Robert: *The Politics of Oil: A Study of Private Power and Democratic Directions*. New York, Macmillan, 1961.

Feis, Herbert: *Petroleum and American Foreign Policy*. Stanford: Food Research Institute, 1944.

Fischer, Louis: *Oil Imperialism: The International Struggle for Petroleum*. New York: International Publishers, 1926.

Garverick, Linda: *Corrosion in the Petrochemical Industry*. Materials Park, Ohio: ASM International, 2000.

Giddens, Paul H.: *The Birth of the Oil Industry*, The MacMillian Company, New York, 1938.

Gramling, Robert: *Oil on the Edge: Offshore Development, Conflict, Gridlock*. Albany: State University of New York Press, 1996.

Hyne, Norman J.: *Nontechnical Guide to Petroleum Geology, Exploration, Drilling, and Production.* PennWell Corporation, Delhi, 2001.

Jentleson, Bruce: *Pipeline Politics: The Complex Political Economy of East-West Energy Trade.* lthaca: Cornell University Press, 1986.

Kent, Marian: *Oil and Empire: British Policy and Mesopotamian Oil, 1900-1920.* London: Macmillan, 1976.

Klare, Michael: *Blood and Oil: The Dangers and Consequences of America's Growing Petroleum Diplomacy.* Metropolitan Books/ Henry Holt, 2004.

Lieber, Robert J.: *Oil and the Middle East War: Europe in the Energy Crisis.* Cambridge: Center for International Affairs, 1976.

Lubell, Harold: *Middle East Oil Crises and Western Europe's Energy Supplies.* Baltimore: Johns Hopkins University Press, 1963.

Marsh, Steve: *Anglo-American Relations and Cold War Oil: Crisis in Iran.* New York: Palgrave MacMillan, 2003.

McNaugher, Thomas: *Arms and Oil: US Military Strategy and the Persian Gulf.* Brookings Institution, Londong, 1985.

Naderpour, N and F Hematti Kordmahaleh: *Petroleum Geosciences*, SBS Pub, Delhi, 2009.

Olien, Diana Davids: *Oil in Texas: The Gusher Age, 1895–1945.* Austin: University of Texas Press, 2002.

Penrose, Edith, and E. F. Penrose: *Iraq: International Relations and National Development.* London: Ernest Bern, 1978.

Philip, George: *Oil and Politics in Latin America: Nationalist Movements and State Companies.* Cambridge University Press, Cambridge, 1982.

Randall, Stephen J. *United States Foreign Oil Policy, 1919–1948: For Profits and Security.* Kingston, Ontario, 1985.

Rister, Carl Coke: *Oil! Titan of the Southwest.* Norman: University of Oklahoma Press, 1949.

Sampson, Anthony: *The Seven Sisters: The Great Oil Companies and the World They Shaped.* New York, 1975.

Simmons, Matthew R.: *Twlight in the Desert: The Coming Saudi Oil Shock and the World Economy.* 2005

Tribedi, S. : *Proceedings of National Convention of Environment in India, Challenge for the 21st Century*, India, 1994.

Vervalin, Charles H.: *Fire Protection Manual for Hydrocarbon Processing Plants.* Houston: Gulf Publishing Co, 1985.

Watkins, Eric: *The Middle Eastern Environment*, London, The British Society for Middle Eastern Studie, 1995.

Wendel, Charles H.: *The Allis-Chalmers Story.* Sarasota, Florida: Crestline Publishing Co., 1988.

Yergin, Daniel: *The Prize: The Epic Quest for Oil, Money, and Power.* New York: Simon and Schuster, 1991.

Index

A

Acidity, 18, 19.
Air Gun, 212, 213.
Annular Safety Valve, 119, 225.
Archaeology, 179, 182, 192.
Artificial Lift, 119, 129, 177, 178, 209, 210, 211, 212, 219.

B

Barefoot Completion, 115.
Basin Analysis, 223, 225.
Bespoke Software, 278.
Breast Cancer, 28.

C

Caesium Vapour Magnetometer, 191.
Caliper Log, 164.
Centralizer, 120.
Chemical Warfare, 52, 77.
Coal Exploration, 183.
Cumene, 15, 16, 17, 20, 23, 93.

D

Data Storage, 49.
Degradation, 7, 11, 38, 51, 52, 83.
Density Logging, 162.
Directional Drilling, 134, 135, 137, 139, 149, 168, 184.
DNA Methylation, 29.
Down-hole, 141, 203.
Dual Porosity, 237.

E

Earthquakes, 77, 79, 82, 181, 228.
Electrical Conductivity, 155, 177, 228, 283, 284.
Exploration Stage, 225.

F

Fabric, 7, 14, 49, 51, 197, 232.
Foams, 9, 11, 13, 72.
Formation Isolation Valve, 120.

G

Gas Lift, 119, 128, 206, 210, 277.
Geological Logs, 150, 155.
Geosteering, 138, 169.
Gradiometer, 183, 190, 194.

H

Hydraulic Fracturing, 70, 71, 72, 73, 74, 75, 77, 78, 79, 80, 81, 82, 83, 84, 85, 86, 87, 88, 121, 131, 206, 216.
Hydraulic Pumping, 209.

I

Incineration, 12.
Ineffective Porosity, 237.
Inorganic Chemistry, 54.

L

Laboratory Synthesis, 3.
Landing Nipple, 119.

Lattice Waves, 272.
Lawsuits, 84, 134.
Layering, 283.
Lithology, 150, 156, 163, 164, 165, 167, 220, 227, 228, 230, 231, 233, 234.

M

Macro Porosity, 237.
Material Phase, 269.
Memory Log, 157.
Mud Logging, 138, 141, 156, 164, 165, 167.

N

Negative Pulse, 170.
Neuroblastoma, 28.
Nitrogen Circulation, 121.
Noise Sources, 216.
Notable Gushers, 107.
Nuclear Explosions, 88, 115.

O

Oil Exploration, 106, 186, 223, 280.
Open Hole Completion, 116.
Ore Quality, 227, 228.
Organic Synthesis, 53, 68.

P

Paleomagnetism, 196, 197, 198, 199.
Passive Seismic, 180, 181.
Perforated Joint, 120.
Petroleum Geologist, 195, 196.
Petrophysics, 219, 227, 230.
Phenols, 2, 20, 18, 19, 22.
Phosgene, 47, 48, 52, 53, 54, 55.
Polycarbonate, 18, 20, 23, 24, 30, 34, 36, 39, 40, 41, 42, 43, 47, 48, 49, 50, 51, 53.
Polystyrene, 1, 4, 5, 6, 7, 8, 9, 10, 11, 12, 13, 14, 15, 23, 270.
Pumpjack, 202, 203, 204.

R

Reactions, 19, 23, 53, 55, 56, 57, 67, 103, 152, 198, 199, 231.
Reservoir Engineering, 216, 219, 224.
Retrievable Tools, 172.
Rod Pumps, 210.

S

Sliding Sleeve, 116, 119, 130, 207.
Spacecraft, 182, 186.
Squeeze Job, 132.
Styrene, 1, 2, 3, 4, 5, 6, 7, 8, 9, 10, 11, 12, 13, 23, 270.

T

Tensor Generalization, 260.
Texture, 163, 221, 230, 231, 232, 233, 236, 283.
Tubing Hanger, 118, 122, 125.

U

Underbalanced Drilling, 147, 148, 149, 170, 171.
Underground Blowouts, 111, 112.

V

Vector Magnetometers, 187, 193.
Void Fraction, 233.
Vuggy Porosity, 237.

W

Water Content, 239, 240, 241, 242, 243, 244.
Water Well Pumpjacks, 204.
Well Logging, 143, 150, 152, 154, 155, 158, 161, 162, 164, 173, 177, 252.
Wellhead, 107, 111, 112, 118, 121, 122, 123, 125, 126, 128, 136, 171, 204.
Wired Drill Pipe, 170, 171.
Wireline log, 156, 157, 169, 229.

❑❑❑